SOLID FLUID BIOTIC
Changing
Alpine Landscapes

SOLID FLUID BIOTIC
Changing
Alpine Landscapes

Edited by Thomas Kissling

VOGT Landscape Architects

ETH Zurich
Department of Architecture (D-ARCH)
Network City and Landscape (NSL)
Institute of Landscape and Urban Studies (LUS)
Chair of Günther Vogt

Lars Müller Publishers

Biotic

The Alps. An Organism

Switzerland – a country whose national cohesion is disputed by nobody. […] However, [the Swiss] have one mass symbol in common, which is never out of their sight and unrivalled like that of no other nation: the mountains. […] To the Swiss, their inaccessibility as well as their solidity symbolize security. And while their peaks above are separate, they are welded together into a single, huge mass below. […] Their mountains are standing proud and solid; they only have to know them well.

Elias Canetti, *Masse und Macht,* Fischer Taschenbuch Verlag, Frankfurt am Main, 2010, pp. 204–205.

Permanence and Dynamics

In his anthropological consideration of mass and power (*Masse und Macht)*, Elias Canetti describes the mountains as the mass symbol of Switzerland. As he noted at the beginning of the chapter on the essence of nations, he is concerned with a reduction to very simple and general features. The description evokes the core of our traditional image of the Alps as a mineral, static, and rigid monolith – as both resistance and barrier.

In fact, the Alps are not a stable structure, but a dynamic and sensitive organism which, in the middle of Central Europe, connects the neighboring landscapes and cultures like a "bridge." Already around 85,000 years ago, human hunter-gatherers probably crossed this imposing mountain region on the tracks of game. Some time later, farmers and their cattle followed in their footsteps. Over centuries and millennia, a fine network of paths and roads was, thus, continuously inscribed into the terrain –

always given the premise of the least necessary intervention in the nature encountered and with the materials and tools available on site.

The crossings from north to south and east to west made possible by this, show up as traces in the local landscape and culture, which – for a long time – was shaped and affected by this "interweaving" of things found with skills, know-how, and civilizing products from more remote places. In this process, the mountains benefited from their proximity to the most important early habitats around the Alpine arc.

Difference

When we travel through the Alps, we encounter a wealth of varied spatial dimensions and qualities. On the one hand, the most diverse forms of life react to the finely crafted relief of their space – a result of the geomorphological processes that have been going on for millions of years. On the other hand, they react to the continually changing external conditions by adapting or migrating. Everything is in motion – *panta rhei* or *everything flows*. This creates a complex fabric that never rests, in which water is the protagonist. In its frozen and solid, liquid, or gaseous aggregate state, the element itself turns into a driver of change. It flows through riverbeds, seeps through layers of rock, drips from the ceilings of subterranean caves or penetrates rock crevices and breaks apart the rock because its volume expands as it freezes. But as an *elixir,* water is also the very basis of life for plants, animals, and humans. As such, its sphere of influence is not limited to the Alps along their topographical border. As a water tower that stores considerable volumes at high altitudes and, thereupon, supplies the valleys and forelands according to the law of gravity, the Alps play a crucial part in supplying the surrounding landscapes. These, in turn, are the result of glacial and hydrological processes. Alpine glaciers and rivers shaped and continue to shape the terrain and often formed the morphological foundations that informed the design of our settlements and cities. These laws prevail to this very day.

Fragmentation

In the last 2,000 years, many of the early settlement areas around the Alpine arc have developed into large-scale metropolitan regions. The resulting expansion of infrastructures had far-reaching impacts on the mountains. By means of expressways and railroad lines, through tunnels and over long bridges, large distances and topographical obstacles can be overcome within a very short time. The infrastructures hereby not only partly ignore the landscapes but also marginalize their resistance. This changes the geography to a substantial extent. The metropolises extend their spheres of influence from the edge along the easily accessible valleys far into the Alpine region and degrade the areas appropriated in this manner to trivial secondary spaces. Tourist destinations are suffering a transformation into urban crystallization foci of the leisure industry and are being reinvented time and again. Meanwhile, the process additionally drains resources from regions that are higher up or inconveniently located in terms of transportation. This causes a fragmentation of the space but, at the same time, reveals new potentials, for the retreat of humans from an area creates space for numerous species. The wilderness begins to reestablish itself, and plants and animals reconquer their traditional habitats – slowly but surely.

Destabilization and New Landscapes

However, man's influence has long since ceased to be merely limited to direct and immediate interventions in the Alpine landscape. Meanwhile, the very driving force of change, *homo sapiens*, has itself mutated into a force of nature. This is not only reflected by the fact that we move more earth, stones, and sand worldwide than all geomorphological processes put together, with the effects of this shifting demonstrated to us in the Alps in a particularly significant way. How far-reaching the consequences of our actions really are can above all be seen in the fact that man-made greenhouse gas emissions are leading to a global warming of the climate. The Alpine region is particularly affected by this. The annual mean temperature has increased by more than 2°C since the middle of the nineteenth century. This is almost twice the global average. But that is only the beginning of this development. Warming is increasing relentlessly and the annual mean temperature will have risen by a further 1.5°C to 5°C by the end of the century. As a consequence, the Alpine landscape is changing quite significantly and at a rapid pace, too: the zero-degree line is moving to higher altitudes and along with it, various plant species that are opening up new habitats. The glaciers are melting and additional lakes are forming in the characteristic depressions of the glacier beds that are becoming ice-free – hundreds of them, in fact, solely in the territory of Switzerland. The permafrost is beginning to thaw. This leads to a precarious destabilization of the rock, which results in an increased threat of natural hazards such as rockfalls, debris flows, or rock slides. The beautiful and sublime Alps, which have stood solid and immovable for thousands of years, thus reveal their innate fragility.

Mutualism
The transformation of the Alps is accelerating. However, in a global context and with the involvement of the local population, this process also holds a variety of potentials. Endowed with numerous resources (renewable energies, fresh water, or a high level of biodiversity), the Alps will continue to develop into an "ecological island" in the middle of Central Europe given the premise of an increasing "solidification" of the urban periphery. However, "within" and "without" are not to be regarded as separate spheres whose relationship is at best one-sided. Rather, they should be seen as spaces which, in the future, should be even more strongly interconnected in a symbiotic way – without assimilating each other in the process though.

In the course of this development, the importance of the Alps will continue to increase. Any conflicts of interest and use that already exist today are likely to become even more acute in the future. And any discourse about the future of the Alps, focused on the assertion of particular interests with a view to an individual case, or thinking along merely administrative lines, will not be very effective. What is needed is a superordinate view. The existing landscape should be placed at the beginning of any such considerations, with the aim of forming a basic consensus regarding possible developments in the area. Such a "profile of Alpine landscapes" in all its variability could provide a basis for further action. The fallacy that this is a matter of reproducing traditional ideas should be resolutely resisted and attention should be paid to the active creation of new images and definitions. Always aware, of course, that a landscape is not something static, but a complex structure that is constantly in the process of becoming.

How Will We Live Together?

Against this backdrop, the contributions by VOGT Landscape Architects and the Chair of Günther Vogt at ETH Zurich address the Alps and their dynamics with a view to a broader context at the 17th International Architecture Exhibition in Venice. This book documents the works exhibited while creating an extensive space of reference along the motifs of the exhibition posts entitled *Solid, Fluid, Biotic*. The discussion includes experts from disciplines near and far, who are shedding light on the themes from their diverse perspectives by means of scholarly essays, art interventions, and field trips.

Thomas Kissling

>> Already about twenty years after Leonard von Matt's photograph (on the right), modernity made its way into the valley. The dam on the Göscheneralp was completed in 1960 as a replacement for an unrealized reservoir in the Urseren Valley. The former settlement was flooded by the lake. Farm buildings, residential houses, and the church had to be demolished and the entire population had to be forcibly relocated. The huge dam structure was built as a planted earth dam (rockfill dam). 9.3 million cubic meters of stone, rock, and earth were moved and piled up for this purpose. The reservoir thus created is used for the production of electricity and supplies the Göscheneralp-Göschenen stage of the Göschenen power plant with water.

pp. 12/13: Chelen Glacier, 1981
p. 14 top: dam from the lake side, view to the northeast, 1960
p. 14 bottom: view of dam and dam chain from the Hinter Bründli material extraction area, 1959
p. 15: Göscheneralp Lake, Göschener Reuss, 1964
pp. 16/17: Göscheneralp Reservoir, 1991

Leonard von Matt, *Potato Field on a Granite Stone,* Göscheneralp, Canton of Uri, 1940s

The potato field on a granite rock on Göscheneralp was established with great effort. After the humus layer was applied by hand, the farmer's wife tended her vegetable garden with a lot of attention and care, because the potato, with its far greater yield than the usual crops at the time, such as corn, was part of a diversification strategy and, therefore, an inherent part of the agrarian culture. Crop failures due to inclement weather or as a result of disease could thus be cushioned and hunger at least partially alleviated. Nevertheless, for a long time, farmers were forbidden to cultivate the novel plant from overseas on the commons. At that time, the potato was still considered pigswill (or rather, lousy grub) and, therefore, could not be cultivated in the vegetable garden common to all.

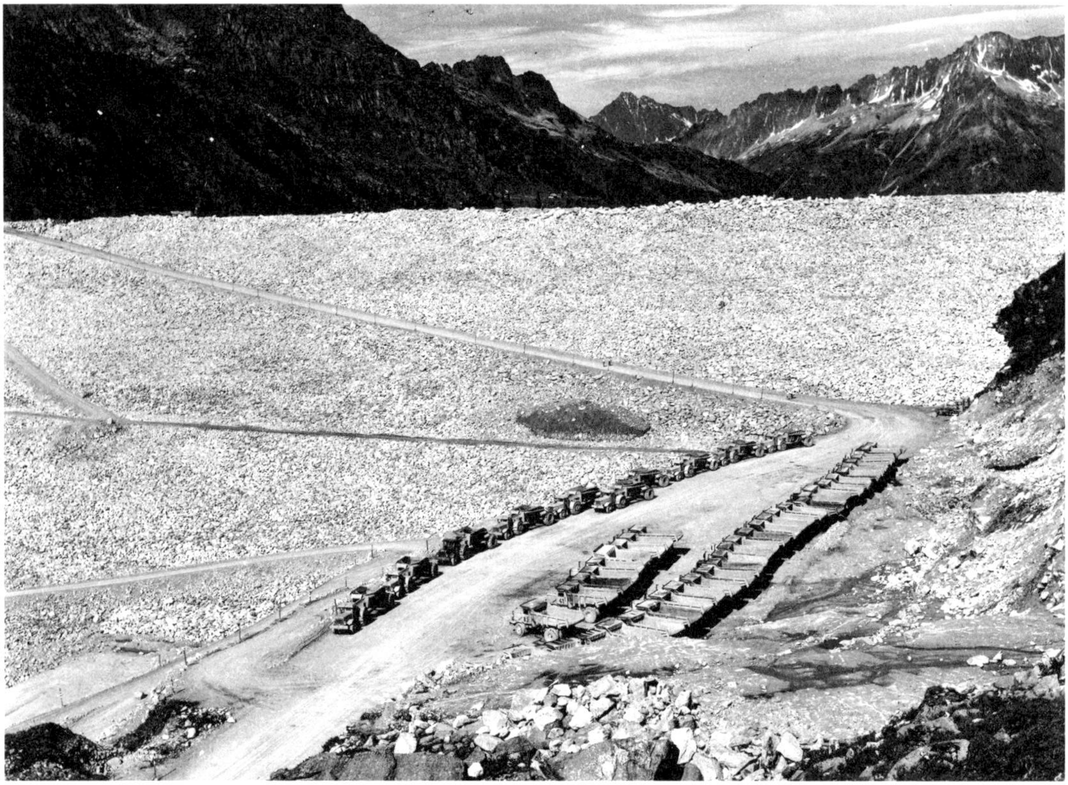

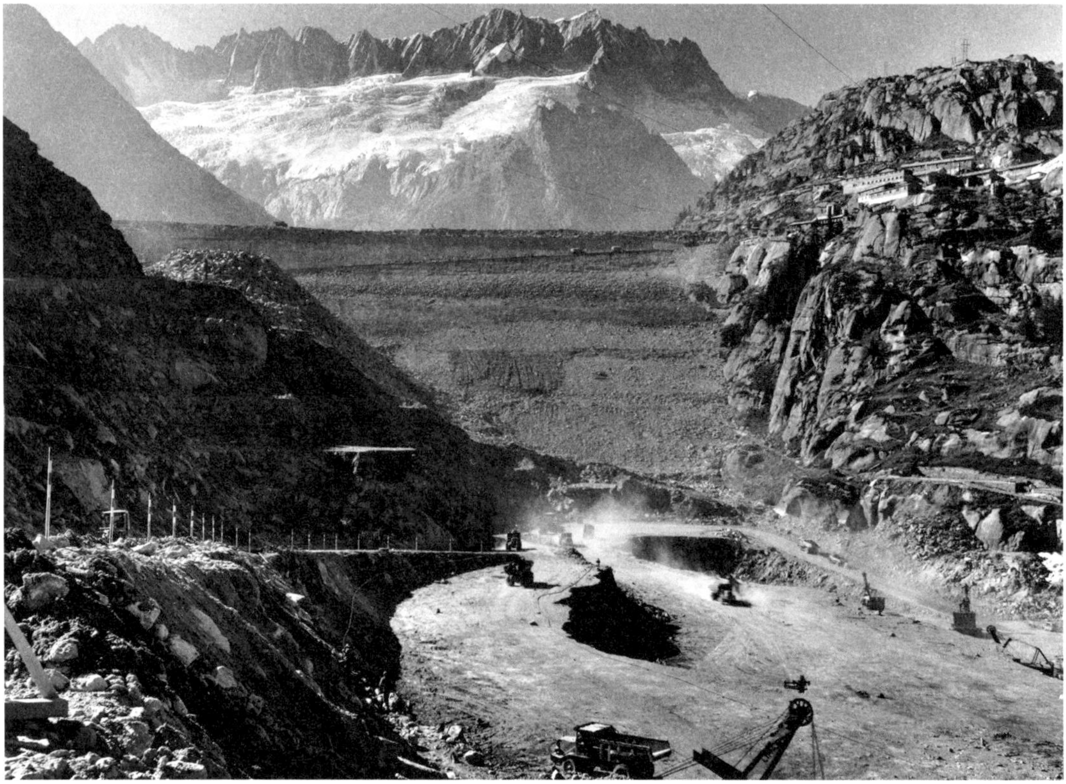

Becoming Aware of Alpine Dynamics – While Hiking

Markus Ritter

A hike in the Alps quickly leads me to a remote area, which I traverse – solitarily and boldly. The beguiling, spicy scents of the Alpine flora, the soothing freshness of bubbling brooks and rivulets, the shrill whistle of marmots and the buzzing call of the spotted nutcracker – all of these fail to captivate my senses though. My whole attention is taken up by the space itself. My look creates both order and overview, structuring the vista in front of me from the foreground to the most remote distance. Staggered and structured into different altitude zones, I am encountering spaces in these mountains whose very diversity is exciting. And as always, motion, flow, and dynamics are to be discovered in these Alpine spaces, too.

Hiking on the Paths of Utopia

I never go for a hike completely unprepared. Especially not in a complex landscape such as the Alps. In the mountains, in addition to the usual hiking preparation and equipment, site-specific documentation, and reliable maps, there is a lot more to think of. An inner picture of the overall spatial conditions accompanies me on each hike and provides some orientation. I have already formed an idea of the destinations, routes, water conditions, and the flora and fauna beforehand. While hiking, I now focus on weather conditions, mountain air, rocks and minerals, tectonics and the emergence of mountain formations, erosion, water, climate and climate change, the traces of glaciers, the history of vegetation, flora and fauna, the management of the cultural landscape, the road and path network and traffic, statistics of emigration and out-migration, tourism, mountain sports, a spontaneous colonization by scrub, and the wilderness.

Landscapes in a standard look, from the series of *Mittlere Bilder [Central Pictures]* in the *Bilder, Ansichten [Pictures, Views]* series by Peter Fischli / David Weiss: *Untitled*, 1990/97. © Peter Fischli and Oskar Weiss

In the mountains, the success of a hike depends on trained vision: exploring the space in front of you, discovering and interpreting its signs and, basically, beginning to understand the landscape. Everything I experience on such a hike, I imagine being both dynamic and transient. To understand it, I shall have to get acquainted with an abundance of small-scale and highly varied facts.

A *strollological* walk, to paraphrase Lucius Burckhardt, the inventor of the term, is the tried and tested research medium to do so. Let's set out on the path then! And let's see what we shall encounter on our walk through this Alpine cabinet of wonders.

A Small World

Strollers tend to be contemplative – they seek to understand the Alpine space in all its diversity and difference. They collect narratives from all kinds of sources and people. And there are such wonderful narratives about Alpine landscapes! Madame de Sévigné was the first to add to its literature with her lively letters. From Grignan Castle near Montélimar, she looked across Languedoc and Provence to Mont Ventoux and wrote to her daughter: "J'aime fort tous ces amphithéatres! [I simply love these amphitheaters]." In her *Correspondances* written between 1669 and 1696, she never tired of describing the snowy mountains as "ces épouventables beautés [these terrible beauties]." The immediate freshness of the Marquise's perception and narration opened up an epochal new view of the Alps. Thus, the literary and pictorial exploration of the Alps began, though quite hesitantly at first.

When strolling or hiking through the Alpine mountains, one is struck by the pronouncedly small-scale nature of the landscape. Father Placidus a Spescha saw in the Alps nothing but "valleys whose flanks are the mountains" (1818). This archaic image applies to the entire Alpine arc, from the Ligurian and Provencal Maritime Alps to the Leitha Mountains in the Vienna Basin. Each valley, enclosed by high and steep slopes, has its own peculiarities. The division of the landscape is continued by the political allocation of space. The actual use of the space is however determined by the self-image of the mountain

Caspar David Friedrich, *Frau vor unter-gehender Sonne* (also: *Sonnenuntergang, Sonnenaufgang, Frau in der Morgen-sonne, Morgenlicht) [Woman in Front of Setting Sun (also: Sunset, Sunrise, Woman in Morning Sun, Morning Light)]*, ca. 1820. Folkwang Museum Essen

population, whose products and services change with the valley communities. Scattered regionalism is very pronounced, with autonomous and semi-autonomous territories such as Monaco, Liechtenstein, the 15 Swiss mountain cantons, Bolzano, and Trentino.

Overcoming Isolation

Valleys are actual communities. Common property is widespread in Alpine primary production. The negotiation of rights of use takes place co-operatively, according to rules established by the community. In fact, in the Alpine regions private ownership is not conducive to a competi-tive and sustainable use of natural resources.

The valley community way of life has a culturally and sociologically exclusionary effect. In order to alleviate it, many different things have been initiated everywhere. The endeavor to overcome isolation is constant and omnipresent; thousands of passes that can be traversed on foot allow encounters with neighbors from any part of the landscape. Roads lead out of the valley into the river valleys and to the urban and commercial centres with which the Alpine region forms a larger economic and geo-graphical entity.

Trail networks open up adjacent regions and communities. Routes through the Alps are gradually inscribed into the landscape. The paving of the ground from paths to trails to roads expresses the purposeful allocation of path networks, with destinations and means of transport the decisive principle.

Mid-summer postbus ride on the Flüela Pass between Davos and Süs (today called Susch). View into Val Grialetsch from the Chantsura hairpin bends, Johann Feuerstein, *Susch-Davos Post-bus (near Susch)*, ca. 1935, Scuol. Photo: Feuerstein © Fundaziun Fotografia Feuerstein

Emil Rittmeyer and Wilhelm Georgy, *Runsen und Ungewitter [Barrancas and Thunderstorms]*, genre painting, in: Friedrich von Tschudi, *Das Tierleben der Alpenwelt. Naturansichten und Tierzeichnungen aus dem schweizerischen Gebirge,* 1868, 8th ed., Verlagsbuchhandlung J. J. Weber, Leipzig, 1890. © ETH Library, Zurich

Where the Alpine arc stands like a crossbar in the way of long-distance relations, base tunnels nowadays exclude entire mountain ranges on the route to one's destination. However, the metropolises in the Alpine foothills should also be given rapid access to the contrasting Alpine landscapes. Busyness is expressed by all these facilities, while short-term stays of sports and pleasure tourists do not allow to let time pass by idly.

Catastrophism

Mankind has always been fond of imagining the primeval era as a rather dull and sinister world. The words of Moses, the prophet, as translated by Buber and Rosenzweig are:

The Earth was madness and confusion.
Darkness reigned over the face of the primeval vortex.
God's roar flying over the waters' surface.

That Earth changed its appearance, climate, and inhabitants several times, is part of the popular imagination and narrative. Echoes of this legend are told by the poets of the Golden Age, images of paradise present it, and Plato mentions it in his descriptions of cosmic renewals. That certain cultures retreated as a result of a gradual change in temperature was a commonly held popular opinion.

Catastrophic images accompany the ideas of the formation and the experience of the disintegration of the Alps. The explanations of the folding up of the Alps fit in with this type of "catastrophism." The positional relationships of rocks in the overall structure of the mountains were already noticed by the ancients. But around 1820, the change in the Earth's surface, the dynamics of mountain folding, finally became a focus of geology. Volcanism became widely recognized. And the uplifting of the (stratified) mountains by tumultuous processes of uplifting and the subsidence of ancient rock deposits entered people's imagination. Goethe learned of this in his old age, opposed it and temperamentally dictated to his scribe John: "The matter may be as it wills, so it must stand written that I curse this accursed *Polterkammer* [poltergeist's chamber] of new world creation […]." It was inconceivable to the

H.C. Escher von der Linth, *Die Glarner Überschiebung am Martinsloch* *[The Glarus Overthrust at Martinsloch]*, July 22, 1812. H.C. Escher estate. Graphische Sammlung ETH Zurich

old man that the Himalayan mountains were lifted out of the ground at an altitude of 9,000 metres and yet rise so rigidly and proudly into the sky as if nothing had ever happened. In order not to delve deeper into this, he simple imputed a case of talented rhetoric to his friend Alexander von Humboldt, who had presented this view in Berlin: "Every oral presentation wants to persuade and make the listener believe and convince him. Few people are capable of being convinced; most can be persuaded though.…"

The disturbing image of a *Polterkammer* has since given way to the even more monumental description of gigantic movements of the Earth's crust. The Alpine peaks are not heaps of rubble but folds of the Earth's crust, comparable to the shapes of a cloth lying flat and then pushed together to form folds. Thrust forces and, as their counterpart, the old granite and gneiss massifs are the cause of the horizontal displacement of entire rock formations and their folding. Discoverer Alfred Escher only dared to record this idea, indeed this vision of a "Glarus double fold" in his notebooks, because: "No one would believe it, they would take me for a fool." But in 1884, Parisian geologist Marcel Bertrand dared to imagine this mightiest of things and would soon find widespread approval. The Glarus "double fold," he said, was a single horizontally folded "blanket" that had been displaced 40 kilometers above the bedrock! It is the first example of a previously unthinkable "overthrust" as it has since been widely demonstrated. Around 1900, the "blanket theory" prevailed against an eloquent opposition that exclaimed: Impossible! Nonsensical! Unthinkable! But a courageous generation of Alpine geologists recognized in ever more in-depth studies that the Alps are much more crumpled, kneaded, and tangled in their folds and overthrusts than anyone had dared to imagine a comparably short time before. The movement of advancing blankets or layers until the final *mise-en-place* is thoroughly exciting though.

Thinking the unthinkable led Alpine geology to the accurate insight. "We are in the most productive creative period," exclaimed Albert Heim delightedly (1922). "Like scales falling from our eyes." A tremendous perspective of new and beautiful tasks, new points of view, and new ideas had brought geology to fruition. The revolutionary view, which Goethe defiantly shrank from, took hold at last.

Alpine Weather and Climate

In the mountains, weather often changes rather abruptly. In the valley, hikers are not aware of the approaching weather fronts. But the butterflies of the High Alps quickly react to such weather events. *Parnassius phoebus*, the small Apollo, only flies in bright sunshine. The same as the umber Erebia and other butterflies, it will immediately settle on the ground and seek shelter underneath the leaf canopy once the sun disappears behind a cloud. Where there was a lively dance of butterflies a short while ago, there is all at once not a single one of these sun-loving creatures to be seen. Hikers are well advised to pay attention, when suddenly, no butterfly is fluttering around anymore.

The Alpine climate markedly differs from that of the foothills and plains. It also varies quite amazingly from region to region. It is generally wet and cool, but in the rain shadow of the mighty mountain ranges, thirteen exceedingly dry valleys stand out for their local climate. They are located inside the Alpine arc, in sun-drenched valleys from the French Alps in the South-West to Carinthia in Austria. These valleys are privileged in terms of warm temperatures and climate as the duration of sunshine here exceeds 2,000 hours per year, the annual precipitation is a very low 500 to 800 mm per year, and in summer there is an average of barely 170 mm of rain. The flora and vegetation of the dry inner Alpine valleys is much enriched by Mediterranean elements and species from the Pannonian steppes. These quite isolated plant occurrences distributed all over the Alpine arc are presumably the pioneer populations once dry summers and an increased mean temperature start to prevail in the Alpine foothills.

Seeing the Climate Change with Your Own Eyes

250 years ago, glacier research began as a meticulous description of the actual situation – but as a snapshot, it simply melted away. Around 1770, meteorology discovered "climate" apart from weather, which was experienced as eternally changeable, but only gradually understood

James Lee Byars, *A Drop of Black Perfume*, July 24, 1983, postcard, performance on Furka Pass.
Photo: Balthasar Burkhard, FurkArt
© Furkablick Institute / Alfred Richterich Foundation

in more detail. This evolution of ideas within a mere 250 years is quite amazing.

In the eighteenth century, curious people increasingly climbed the seemingly impassable high mountains. Pierre Martel, an enlightened engineer and geographer, noticed large boulders lying around on his way home from the Mer de Glace [the Ice Sea] to Chamonix: no simple rockslide, he quickly realized, which somewhat irritated him. With a smile, the farmers in the valley pointed out that this was rock debris carried into the valley by the Glacier du Bois! The peasants in the Savoy Valley had long since recognized the link between glaciers and erratic boulders.

In the summer of 1787, natural philosopher Horace Bénédict de Saussure, a native of Geneva, his servant, and 18 companions reached the summit of Mont Blanc. The year before, his assistant had scouted out the route to the summit for him. The convivial hike is famous as de Saussure's first ascent and proved scientifically rewarding. The result was the realization of the connection between glacier level and climate. From then on, global warming was considered the cause of glacier melt.

In 1815, Alpine herdsman and chamois hunter Jean-Pierre Perraudin told scholars of his vision of a total glaciation that reached up to great heights. In 1837, Karl Friedrich Schimper coined the term of *Ice Age* in his humorous, twenty-two-stanza *Ode to the Ice Age*. It is he, who, together with Louis Agassiz, formulated the first great synthesis into an ice-age theory, stating that glaciers were simultaneously plough, file, and sledge and spread Alpine debris and rubble in ground and wall moraines (angular material of the lateral and medial moraines) far across the mountainous area in a spreading, landscape-shaping pattern.

Count Gaston de Saporta, who dedicated his entire life to science, came from an old, Aix-en-Provence family. A Frenchman and Darwinist, there was no other like him at the Académie at that time. Accordingly, his texts are exciting and well worth reading. He describes the "old climates" in his book *La Végétation du Globe dans les temps antérieurs à l'homme [The Flora Before the Appear-*

ance of Man] (1879). He recognized the way heat and water are distributed on the Earth's surface as key for entire climatic eras of the Earth. "One must study the climate if one is to give an account of the conditions that produce, stimulate, sustain, or weaken life." These statements herald the great theme of climate change.

The fact that climate conditions are not eternally constant, but on the contrary surprisingly dynamic and fragile, is currently concerning us in today's phase of a particularly accelerated climate change. In the Alpine region, the signs are unmistakable. The high mountains are something of a climate island on the European continent, comparable to the Arctic north. The Arctic-Alpine distribution pattern of numerous plant species is proof of this climatic analogy. More than 90 grasses and herbs are common to the two far-flung landscapes, and an attentive hiker regularly comes across plant species with an Arctic-Alpine distribution in the Alps.

But how is climate change affecting the Alps? The snow cover is decreasing, the 0°C limit is higher every year, glacier melt is accelerating, heat days in summer are becoming much more frequent, and the annual mean temperature rises 1 or 2 degrees. This becomes noticeable in the flora of the mountain peaks. Botanist Josias Braun, a native of the Grisons, climbed the mountain peaks of his homeland around 1900–1912 and described plant life at its outermost limits. Veronika Stöckli studied the flora of the same peaks 100 years later and found an upwards migration of high mountain plants. On the summits, both the number of individuals and the number of species

Otto Morach, *Der Weg zur Kraft u. Gesundheit führt über Davos [The Way to Strength & Health Leads Via Davos]*, 1926. Poster for the Davos Tourist Office © Museum für Gestaltung Zürich, ZHdK, Plakatsammlung

have increased. The ascent so far covers 80 to 100 meters in altitude. Animals living in the high mountains are also attracted to the heights. The ptarmigan is now quite common 300 meters higher up. However, the upward migration is a flight to nowhere if climate warming continues.

The Alps as a Refugial Space – A Therapeutic Stay in the Mountains

Although Hippocrates and Galenius had already recognized and praised the therapeutic effect of the mountain air, the Alpine world remained completely alien to antiquity. Abysses, storms, and robbers were emblematic of the Alpine landscape in antiquity. Leonardo da Vinci and Conrad Gesner, and a little later Johann Jacob Scheuchzer and Albrecht v. Haller, sought out new and positive features in the Alpine region such as spas, exercise, and recreation. Alpine minerals, medicinal herbs, healing waters, milk, and butter in therapeutic treatments became modern signatures of the Alpine mountains.

Medical historians tell us about the year 1854 that Hermann Brehmer, a Silesian physician, already used high-altitude air and treatment in a sanatorium for pulmonary tuberculosis. It is thanks to Brehmer that climatotherapy became known to the urban elites in Europe. In Brehmer's wake, healthy refreshments und healing powers were attributed to the high-altitude climate and mountain hikes. The sanatoria in Davos were the first Alpine resorts of a sophisticated and urban kind.

However, touristic uses of Alpine regions are no less powerful than therapeutic ones. Climbers have "rediscovered" the Alps. They give Father Placidus' "flanks" of the valleys a new and economic significance. Their truly "outdoor" view, as original as it is dangerous, imbues previously useless landscapes with a market value. Ever since the Golden Twenties, new types of equipment have allowed sports to flourish in winter landscapes, too. A foreigner's gaze created an interest in hitherto unused and hazardous zones. Thus, two very different cultural practices were inscribed into the Alpine region in modern times, along with their signs and destinations: tourists and traditional residential ones.

The Invention of the Alps in the Context of Industrial Modernity

In the second half of the nineteenth century, the view of urban elites led to the current differentiation of the Alpine region. By this process, the Alpine space was divided into an extensively and an intensively utilized one, into wilderness and resort spaces. The most momentous of all catastrophic narratives about the wilderness of the Alpine region is Friedrich von Tschudi's popular *Thierleben der Alpen [Alpine Fauna]* (1855). It virtually demonized its large mammals and "birds of prey." The newly awakened industrial society projected its fears about social change into these "predators" and positively feared any physical encounter with large mammals. The Alps have become somewhat edgy since Tschudi's extremely successful text. His *Thierleben* belongs to the genre based on evocations of wild animals that flourished with romanticism and had an extremely tenacious afterlife. "Wilderness" became a black box for festering fantasies and fears.

Discourses about the Alpine region are dominated by landscape heterotopia. Foucault contrasted them with the utopias whose material equivalent they are in the real world. These are spaces of contrariety, whose reality is radically different from conventional places such as a brothel or a sea-going ship with its crew. "Wilderness" is also a space in the structure of heterotopia. If wilderness is protected in a nature reserve, Sigmund Freud compares it to imagination, which is able to mitigate the renunciation of pleasure for the hygiene of the soul: "Everything is allowed to proliferate and grow in it as it wishes, even the useless, even the harmful."

Engadine bear hunters, their prey and the public. Photo: Josef Rauch, *Bärenjagd, der letzte Bär, Scuol [Bear Hunting, the Last Bear, Scuol]*, picture supplement to *Freier Rätier*, 1930, Disentis. © Keystone / Sabine Dreher Collection / Otto Roth

For the Alpine region, the idea of a heterotopic division of space is still retold today with rigid utilitarianism. What urbanity is not able to embrace, is pushed back into the wilderness as a "wasteland," as it were. The disadvantage of this dual system is its great lack of definition. Between urban centrality on the one hand and wilderness on the other one, there is a gaping expanse of space.

The modern process of splitting the Alpine region is exemplified by the Sarasin cousins in Basel and Carl Schröter in Zurich. The former pushed through the creation of the Swiss National Park in the Lower Engadine, the latter the rationalization of mountain farming. The former devised a "national park" with a ban on entering, the latter proved himself as a "patriotic farmer" at university.

Carl Schröter – the "Patriotic Farmer"
Carl Schröter habilitated at the age of 23 and, in 1879, was given a lectureship for special botany at ETH Zurich. On December 19, 1883, he was appointed Professor of Botany. From then on, he also taught Agriculture and Forestry. He developed plant geography into a wide-ranging field of research and taught at ETH Zurich until 1926. His lectures and excursions were extremely popular with the students. His liveliness in describing, explaining, and communicating are legendary.

A basic trait of his nature was "his victorious, indestructible optimism," which predestined him to be a pioneer of nature conservation. Schröter was one of the founders of the Swiss National Park in 1906 and militantly advocated the protection of plants: "The mountains have always played a significant role as 'flora reservoirs,' as refuges, in the history of flora." He disqualified any mass picking as a threat to the Alpine flora: "What we must zealously deny, frown upon, forbid and persecute is […] the prestigious dragging home of giant bouquets, the tasteless decorating of Alpine alpenstocks with flowers, the garnishing of the hat with gentians […]." Botanist Paul Fournier puts it more laconically: "Détruire ce qu'on aime, c'est une assez mauvaise façon d'aimer […]." ["Destroying what you love is a pretty bad way to love […]."]

Schröter represents the rise of industrial agriculture towards the end of the century. In fact, ETH Zurich and the Swiss Federal Administration were leading the way, not nineteenth-century socially organized forces. Around 1890, Carl Schröter and F. G. Stebler published their ground-breaking work on grassland and Alpine farming. Fertilizer trials have as their goal the achieved yield weight on a scale. The patriotic ETH farmers made a bee-line to a more extensive increase of production. One of them, David Guggerli, speaks of a utopian machine that was ramped up at that time.

Paul Sarasin – "the Napoleon of Nature Conservation"
The son of a factory owner, Paul Sarasin grew up in Basel and was interested in the natural sciences. A fateful meeting brought him together with Fritz Sarasin, the son of the mayor and his distant cousin. Joint studies, joint stays abroad, and joint travels characterize the years of their youth. Back in his hometown after 15 years away, the two shared the management of quite some munici-pal institutions. Together, they also founded the popular *Einfrankenverein* in 1909, the One Franc Association as the nature conservation association was called.

The Sarasins had come to know the world as a desecrat-ed organism. That's why they pushed through the creation of a "total reservation" (national park) in the Lower Enga-dine, after the huge project of a world nature reserve for the Arctic had first been on their mind. They proceeded with the help of the Federal Council, whom they visited in person. Paul Sarasin found it difficult to handle any opposition and was not open to even the most well-intentioned advice once he had arrived at a decision. In 1912, he decided on the national-park project. But when business faltered and financing became an issue, Sarasin personally assumed financial responsibility for all outstanding lease payments for the National Park. In 1914, shortly before the outbreak of World War I, the Swiss National Park was finally founded.

In the early industrial age, civilization-weary intellectuals on both sides of the Atlantic divided space into "cultural landscape versus wilderness." Large-scale wilderness

Caspar Wolf, *Le pont et les gorges de la Dala à Loèche, vue en amont [Bridge and Dala Gorge in Leuk, Seen Upstream]*, ca. 1774–1777, oil on canvas, 82.5 cm × 54.2 cm, Musée d'art du Valais, Sion, inv. BA 343 © Valais Cantonal Museums, Sion. Michel Martinez

parks were created, the Yellowstone Park in 1872 in the USA and the "National Park" in Switzerland in 1914. The Swiss National Park, with its ban on access outside a modest network of paths, privatized 143 km^2 of protected area for a scientific elite. Its interdisciplinary goal was to create a wilderness.

Transformation and Metamorphosis

The resident Alpine population did not remain settled in their valley communities. The wave of emigration in the nineteenth century was followed by an out-migration to cities and metropolises.

However, this depopulation did not take place on a massive scale. On the contrary, the population in the Swiss Alpine region has been increasing again for several decades now. This is a true turnover. Locals are still moving away, but new residents are arriving. Not necessarily for residential purposes, but for a part-time use of properties and flats. The villages are filling up again, currently even in an accelerated manner, triggered by the pandemic.

Those valley communities that marketed all their assets in the boom of the sixties to nineties have become cramped and inferior. In some Alpine regions, the transport infrastructure allows time-saving interaction with the metropolitan hinterland. Not long ago, investments in Alpine resorts were a definitive trend. But the "jet set" hardly ever really arrives in the Alpine landscape down in the village and up at the cable-car mountain station. Here, it surrenders, *faute de mieux*, some comfort. It does excuse the larval local form of comfort and convenience for short periods of time, but fails to take the step into the contrasting experiences of an Alpine landscape. In this sense, the Alps remain the backdrop of a hip meeting place.

Where a valley community and municipality can still demonstrate singular qualities in the Alpine region, the future still beckons. A future that does not belong to those who belatedly want to follow urbanization. Overuse creates a sense of deficiency. Compensation is sought in newly founded local economic cycles. Some valley communities are already firmly involved in such an endeavor.

Gilles Monney, *Die Gegenüberstellung des Landschaftsbildes "Dalaschlucht" von Caspar Wolf mit der heutigen Situation [The Juxtaposition of Caspar Wolf's Landscape Painting "Dala Gorge" with Today's Situation]*, 16.07.2014, Dala-Rotten, in: *Caspar Wolf und die ästhetische Eroberung der Natur*, Kunstmuseum Basel (ed.), 2014. Photo: © Gilles Monney

Difference as a Moving and Stabilizing Moment

In the modern era, the control of development processes in the Alpine landscape is gradually taken away from those directly affected. Structurally weak rural areas in the Alps are managed from afar. Farmers and inkeepers manage economic value-added chains that have fallen apart. What used to be united in the eyes of the resident Alpine population has drifted apart into multiple perspectives.

As diverse as the natural features of the valley communities and Alpine region are, so too is its socio-economic diversity. In small-scale settings, the decisive characteristics of language, denomination, poverty, or prosperity as well as rural and urban locations are mixed. The cluster of such a mixture has an identity-forming effect on the communities. In the Alpine region, "the other" is omnipresent in the neighbourhood, as a matter of course.

There is still a lot of ambivalence in the Alpine region. Many landscapes are by no means semiotically filled in and fully interpreted as to their emblematic aspects. There are thousands of so-called "soft factors" lying dormant that could also be exposed as unique selling points. This is the greatest capital that a spatial unit can preserve as its treasure, its potential for exploration.

Diversity stabilizes when it is thoroughly stirred and occasionally remixed, whereupon the diversity of mixes can ultimately be changed. In this case, a kind of identitarian balance of power prevails. Therefore, the mixture also has to develop variance. It will have to remain transitory. And it will have to be able to take up and integrate fashion trends and sociological changes. Tolerance of differences is a primary virtue in the Alpine region.

We plead for an acelerated production of differences. We envisage an orderly retreat from the surface for parts of the Alpine space, instead of the placement of identical things in fixed spatial units. Postmodern Alpine consumer landscapes have a transient market profile. The privately financed hype of the resorts is followed by a cat's meow to the detriment of the public. A way out is definitely needed.

Joseph Mallord William Turner, *The Fall of an Avalanche in the Grisons,* ca.1810, Grisons © London, Tate Gallery Inv. 489

> Roman Signer, *Bergsturz [Landslide],* 1987, wooden construction with cord pull and locking mechanism, stones, 275 × 140 × 425 cm, time: ca. 5 seconds. Zeppelin Museum Friedrichshafen © Roman Signer

Dreaming of Future Alpine Landscapes

Sometimes, while hiking, our thoughts take flight and carry us far away. That's what happened to me. I allowed myself uncensored ideas about future Alpine walking, strolling, and hiking landscapes. But then, I once again mentally reviewed everything that I contemplated in the nine thematic sections above.

The dangers of the settled Alpine space, and they are devastating, lie dormant in its topography. The decay of the Alps is no less colossal than their emergence. Albert Heim says: "Mountains are the remains between the valleys" and, more catastrophically, but with enlightened connoisseurship: "The structure of the Alps is dissolved into ruins."

Degradation had already gone hand in hand with the folding. And it always accelerates when erosion is not inhibited by the withdrawal of stabilizing elements. Today, our concerns are summer droughts, storms, and the thawing of the permafrost. Dry warm periods, as they are now, promote valley formation through unleashed erosion.

"Panta rhei" or everything flows, Heraclitus once postulated. In the Alps, this means quite physically the sliding of topsoil, "soil flow," and the decay of the piled up. There is only one topographical direction of development on the mountain: everything strives downwards, towards the valley. If the ice masses do not heave and churn but flow away, the mountains will erode in the warm-climate mode: the permafrost causes erosion through slumping, sliding, and solifluction. Alpine soils are mechanically unstable on steep slopes with a high relief energy. The turf slides in fragments. The storage capacity of the glaciers disappears and any High-Alpine precipitation immediately flows down into the valley.

What do High-Alpine landscapes look like without ice? What will we find: reflecting lakes, grey deserts, or valleys (re)conquered by vegetation?

Engineers have built up the Alpine landscape with the scale of the century's peak event in mind. They seek

to stabilize the dynamics where they endanger our works, and to secure the habitat against dangers. But only against those dangers that have been recognized. It is to be feared that the force of a warm-period dynamic will set new standards for Alpine danger zones.

We need engineers specializing in decay. We need architects and engineers specializing in ruins. Construction experts, that is, who understand the accelerated erosion of the mountains and are able to organize and control it. They will have to measure the space for the harmless, accidental decay that we must allow for. Deconstruction in a hazard mode is called for. A gentle and wise guidance of the processes of weathering and scouring that are gradually taking over. Let us remember that mountains are but the remnants of a monumental erosional landscape – outcrops between cleared valleys.

Cultural practices ceaselessly change. Sometimes slowly, sometimes rapidly accelerating. We ask: can an Alpine landscape, once it has been abandoned and colonized by scrub, even largely depopulated, leaving historical contexts behind, once again become sublime and beautiful? Post-historic landscapes remain to be developed. They would be those without the organization of the postmodern consumer chains that paste over the *genius loci*. Not the familiar landscapes with firmly assigned social roles from the global construction kit, that is.

Lucius Burckhardt sees a way out of the dead end of customized Alpine landscapes. Certain perspectives, he thinks, can only be conveyed through art without being instructive and hurtful. Fortunately, the abundant variety and wealth of knowledge in many valleys and valley communities fully meet the requirements for a potent landscape. They are quite conceivable, these completely new formats of Alpine landscapes. "How will we live together?" On winding paths, those concerned seek the harmonious mixture of human activity, human leisure, and chance. For each valley, things will be a bit different. And that is precisely the charm of the Alpine landscapes to stroll and hike in.

Erratic Cultivation – The Cultural History of Erratic Boulders in Switzerland

Sophie von Einsiedel

Although the rocks "originate" from the same glacier, all three erratics considered here are peculiarly different – and not only as to their shape and measures. For thousands of years they were not only the source of myths, but were also used in different ways. In the course of several inspections and in cooperation with experts, these rocks are to be measured and mapped both culturally, geographically, and temporally and presented as an integral part of an important cultural landscape.

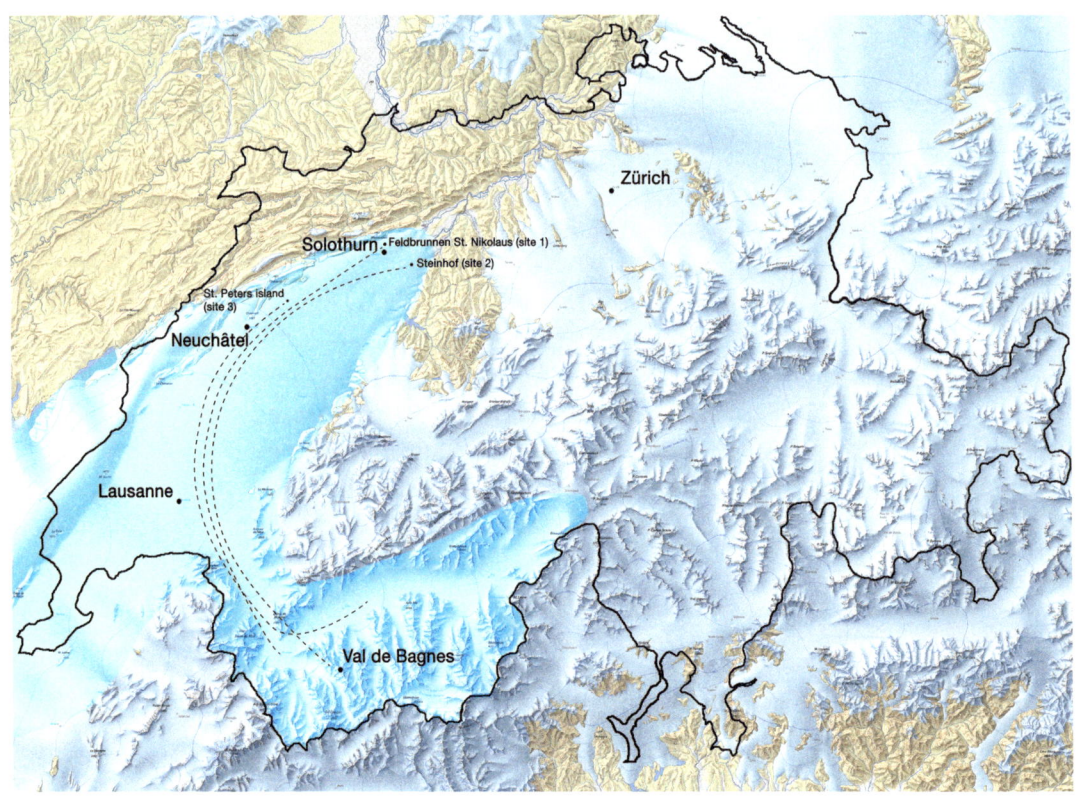

Moved across the landscape by the gigantic glaciers of the last ice age in Europe, imposing erratic boulders once littered the landscapes of the Swiss Central Plateau (*Mittelland*). Since then, many centuries of human settlement around these sites have led to a transformation of these stones into ritual sites, milestones, building materials, and tourist attractions. Inspired by renewed interest in erratic boulders and especially their flora and microbiology, this field trip takes us on a rediscovery of these gentle giants of the geomorphological and cultural landscapes of Switzerland.

The Beginning of Time

Perhaps the most formative force of the Alpine landscape has been the series of ice ages that have repeatedly covered the European continent in kilometer-thick ice sheets during the last 2.6 million years. As part of the last ice age (115,000 to 10,000 years ago) massive glaciers flowed down from the Alpine peaks,

scraping and hollowing out valleys on their way northwards.[1] The glaciers carried with them rocks and debris broken off from peaks or picked up on the way. As they retreated 19,000 years ago, they deposited this material throughout the northern half of Switzerland in the form of moraines and thousands of erratic boulders: rocks that differ geologically from their surroundings; a piece of Alpine rock a long way from home.[2]

⟨ Map of glacial extent around the Alps during the last glacial maximum ca. 25,000 years ago, with annotations showing the field-trip locations, the Rhône glacier and the distances travelled by the *Schildchrott* erratic boulder at Feldbrunnen St. Nikolaus, *Grosser Fluh* at Steinhof and the boulders around St. Peter's Island. Map adapted from © swisstopo

A group photo of an outing to the *Kilchli-flüeli* at Steinhof in May 1946. The stone shows clear signs of repeated sliding, whether for ritual or amusement purposes. Photo © ETH Library, Zurich, Image Archive

Aerial image of Feldbrunnen St. Nikolaus (Canton of Solothurn), with red crosses marking the location of erratic boulders.
Map © swisstopo

Microlandscapes on the *Schildchrott* erratic boulder in Feldbrunnen St. Niklaus (left to right):
1 smooth surface of gneiss;
2 lichen and small clusters of moss on the rock surface;
3 the *Grimmia hartmanii* moss gradually spreading over the stone;
4 lichen and *Pleurozium schreberi*, *Grimmia hartmanii,* and *Paraleucobryum longifolium* competing for space and moisture;
5 the *Pleurozium schreberi* and *Grimmia hartmanii* mosses around a plaque documenting the erratic boulder as a protected stone;
6 *Pleurozium schreberi*, *Grimmia hartmanii,* and *Paraleucobryum longifolium* growing densely intertwined.
Photos © VOGT Landscape Architects

Aerial image of Steinhof (Canton of Solothurn), with red crosses marking the location of erratic boulders.
Map © swisstopo

Microlandscapes on the *Grosse Fluh* erratic boulder in Steinhof (left to right):
1 the Arkesine gneiss of the *Grosse Fluh* has its geological origins in the Val des Bagnes, Canton of Valais;
2 a kerf on a lateral outcrop of the *Grosse Fluh,* where a piece of the stone has been split off and removed from the site;
3 engraving declaring the stone as an object protected by the *Schweizerische*

Naturforschende Gesellschaft [Swiss Society of Natural Sciences];
4 the *Parmelia sensu lato* lichen mingling with the *Pleurozium schreberi* moss;
5 the *Lasallia pustulata* lichen;
6 lichen and mosses such as *Pleurozium schreberi, Grimmia hartmanii,* and *Hedwigia ciliata* compete on the more shaded side of the erratic boulder.
Photos © VOGT Landscape Architects

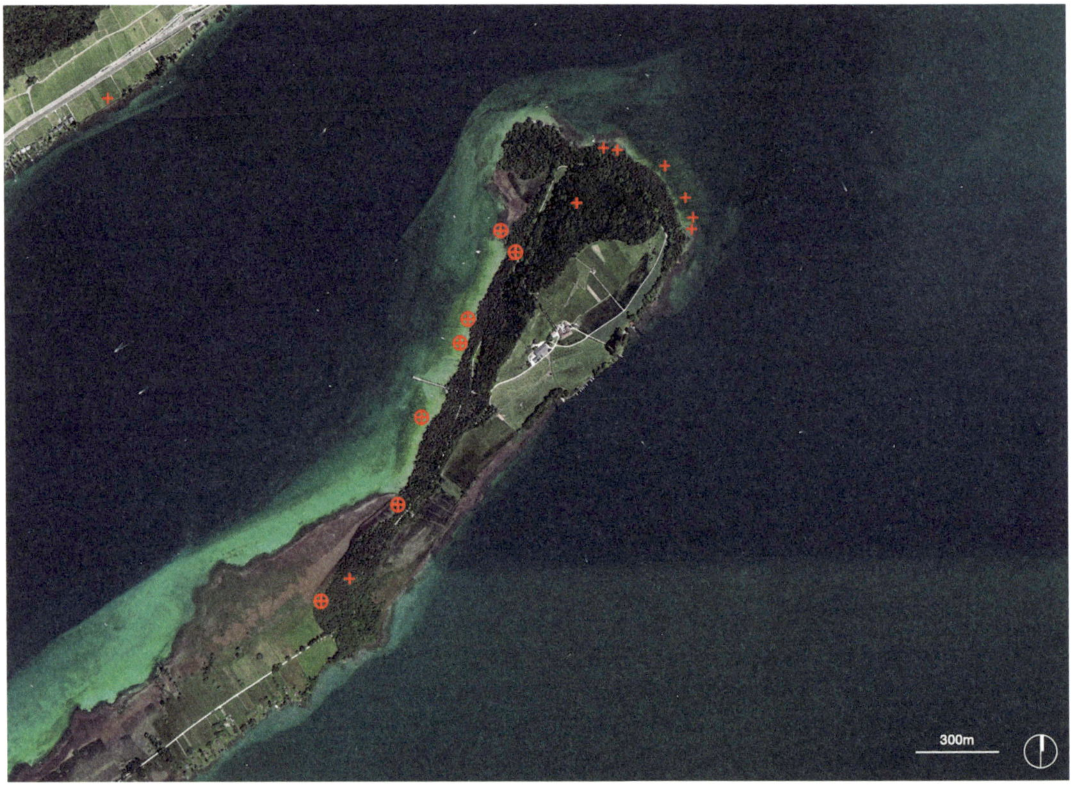

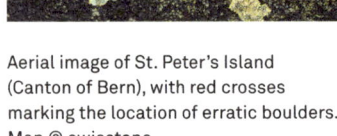
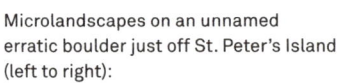

Aerial image of St. Peter's Island (Canton of Bern), with red crosses marking the location of erratic boulders. Map © swisstopo

Microlandscapes on an unnamed erratic boulder just off St. Peter's Island (left to right):
1 quartz and other rock components of the erratic boulder, suspected to be made of Allalinite;
2 lichen and bird guano on the erratic boulder;
3 sun-loving lichen on the highest part of the rock;

4 dense cluster of *Xanthoria parietina* lichen;
5 rings of white crustose lichen dominate one side of the erratic boulder;
6 at the water level of the lake, the rock is encircled by common freshwater algae.
Photos © VOGT Landscape Architects

A Swarm of Rocks

As the ice retreated, *Homo sapiens* spread out from its refugia to repopulate the ice-free landscapes around the Alps. Here, these early humans would have encountered extensive groups of erratic boulders, called block swarms.[3]

This field trip investigated three such swarms in the deep forests of Feldbrunnen St. Nikolaus (Canton of Solothurn), on the open fields of Steinhof (Canton of Solothurn) and at the shores of Lake Bienne on St. Peter's Island (Canton of Bern). The geological formations at these sites were able to withstand the flow of ice above them, forcing the Rhône glacier to deposit some of its load. Initially, the sites would have been densely covered in boulders – as is the case with the forests of Feldbrunnen St. Nikolaus – yet the human cultivation of these landscapes over thousands of years altered the fate of erratic boulders. The field trip revealed the effects of this "erratic cultivation" and the complex cultural, religious, administrative, economic, historic, and national values ascribed to the erratic boulders in each location.

Far from being imposing and static landmarks, the rocks themselves are undergoing constant changes in form, geographical location, uses, and meaning, altering a wider landscape in a shift caused by increasing anthropogenic forces.

Points of Reference

The dynamic journey of erratic boulders through ice advances and retreats during the last – and all previous – ice ages is in stark contrast to their permanent and often monumental presence in the cultural memory of human populations. Not surprisingly then, since the Neolithic (from 9,500 BC), these stones have been used as points of reference for orientation, religious belief systems, cultural practices, nationalistic representations, and scientific discovery.

The earliest examples of human "interactions" with erratic boulders are markings that can be dated to around 4,300 BC.[4] These markings are in the form of cups, worked into the stone by hand, varying in size from 3–30 cm.[5] Such stones, broadly grouped under the term of "cup stones", are prolific amongst erratic boulders in Switzerland. The purpose of the cups remains unknown, with only a variety of theories about their use as containers, sacrificial altars, or astrological and geographical orientation systems.[6] In many cases – such as with cups found on rocks at Feldbrunnen St. Nikolaus and on the *Grosse Fluh* – it is even hard to tell if the cups resulted from human markings or from natural erosion.[7] Time has eroded our knowledge of the meaning of these cups and the complex world views and belief systems they may have represented.

The so called *Chindlisteine* [children's stones] are another example of stone rituals presumably dating from the Neolithic.[8] Until well into the Middle Ages, these erratic boulders were believed to be the origin of children and women would slide down the rocks as a fertility ritual.[9] The *Kilchliflüeli* at Steinhof shows clear signs of frequent sliding – for whatever purpose – and may, therefore, be considered an example of such stones.

The names given to many erratic boulders further illustrate how various cultures and generations integrated these stones into their cultural consciousness. The boulders are named after key Christian practices such as the prayer (*Vaterunserstein*), angels (*Engelstein*), or Easter (*Osterstein*), in an attempt to "Christianize" important sites of competing belief systems.[10] Alternatively, erratic boulders are demonized in name through association with the devil (*Teufelsbürde*), witches (*Hexenstein*), or heathens (*Heidenstein*).[11]

Border Boulder

Almost as old as the erratic boulders themselves is the human desire to know the land, to measure and mark it. As large and unshift-

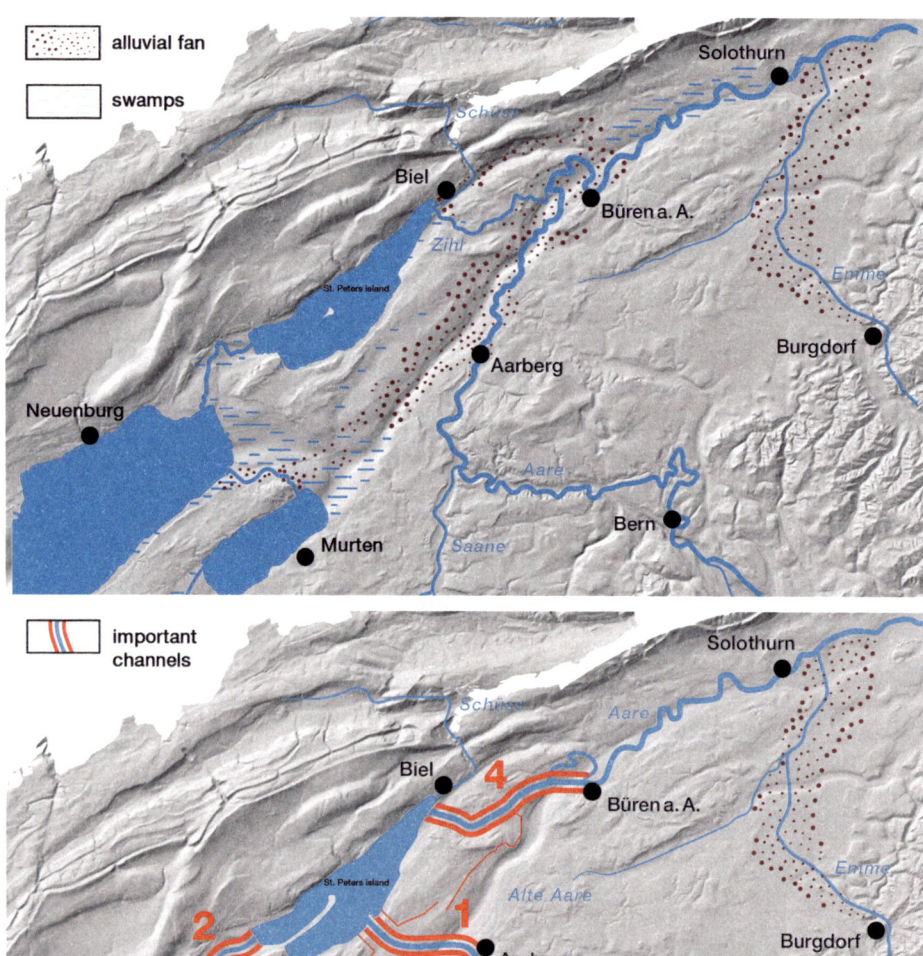

The Riverine landscape along the Aare river before (top) and after (below) the *Juragewässerkorrektur* [Jura water corrections] of 1869–1891. The Hagneck Channel (1) drains Aare waters into Lake Bienne, the Zihl channel (2) and the Broye Channel (3) connect Lake Bienne to Lake Neuchâtel and Lake Murten to balance out water levels. The Nidau-Büren Channel (4) allows Lake Bienne to drain back into the Aare river in a controlled manner. Smaller channels (5) drain the agricultural lands around the lakes. Map adapted from AWA Amt für Wasser und Abfall, Abteilung Gewässer-regulierung, *Regulierung der Jura-randseen Grundlagen und Vorgehen,* Bern, 2012. © VOGT Landscape Architects

able units, erratic boulders were used by the Celts (6th century BC) and later the Romans as milestones as they conquered and surveyed central Europe.[12] For the community of Steinhof, erratic boulders played a crucial role for the etymology of its name – "stone farm" – but also as a point of demarcation to surrounding communities. Steinhof is an enclave of the Catholic Canton of Solothurn, located on top of an end moraine and encircled by the Protestant communities of the Canton of Bern.[13] The Cantonal borders neatly map onto a "ring" of erratic boulders marking the edge of the Steinhof fields. This suggests that these stones have acted as points of demarcation long before official boundary stones were placed alongside them in the eighteenth century.[14]

Walk on Water

Importantly, the vast majority of erratic boulders has no ascribed purpose or meaning. Erratic boulders remain first and foremost compelling evidence of the forces of nature played out at timescales beyond our imagination. Yet, in the last few hundred years, the human "cultivation" of landscapes has become ever more extensive and invasive to the extent that we have become a force of nature in our own right. This is illustrated by the erratic blocks around St. Peter's Island in Lake Bienne, which were only revealed following the *Juragewässerkorrektur* [Jura water corrections] from 1869 to 1891, a large-scale reshaping of the hydrological system

of Lake Bienne, Lake Neuchâtel, Lake Murten, and the Aare river system. The course of the Aare was shifted to flow into Lake Bienne, so the suspended load of river debris would be deposited in the lake basin. This allowed for a better drainage of the surrounding fields which had become increasingly waterlogged since the early eighteenth century. A second channel allowed the waters of Lake Bienne to drain back into the old course of the Aare further downstream, which lowered the water level in the lake by 2.5 m.[15] With near glacial force, human interventions around Lake Bienne brought the erratic boulders at St. Peter's Island to the surface, turning these unnamed and previously unknown rocks into markers of the human domination of the landscape.

A Block of Switzerland

The most persistent "use" of erratic boulders across Switzerland from the Early Middle Ages until the mid-twentieth century was their "mining" for construction materials. The need for hard rocks in lowland areas conveniently coincided with the desire to clear agricultural land of these obstructive boulders.[16] It seemed the obvious choice to work them into fountains or millstones, or break them up into building blocks for churches or gravel for roads.

Erratic boulders in forests such as Feldbrunnen St. Nikolaus were less frequently destroyed, giving us an indication of just

Two historic water troughs carved from granite erratic boulders in Appenzell Innerrhoden. Photo © VOGT Landscape Architects

how densely scattered erratic boulders used to be, and how many of them have been moved or destroyed. Boulder clearances intensified throughout the eighteenth century and spiked further with the advent of the railway, as the rocks were used for the track bed.[17] In 1850, three teachers from Steinhof documented the remaining thirty-nine erratic boulders around the site, though many of them were mined for the rail tracks between Olten and Bern between 1855 and 1857. Around then, erratic boulders began to catch the attention of the Swiss scientific community as they played a central role in the development of the ice-age theory in the early nineteenth century.[18] The theory of an ice age covering large parts of Europe and moving enormous boulders was in fact developed by numerous European scholars, but in Switzerland these discoveries coincided with the foundation of the Swiss Confederation in 1848.[19] The research around erratic boulders and their protection became a nationally unifying endeavor and, in 1867, professor of geology Alphonse Favre published *Appel aux Suisses* [Appeal to the Swiss] urging everyone to protect erratic boulders for the sake of national pride as much as scientific research.[20]

Following this plea, in 1869 the *Schweizerische Naturforschende Gesellschaft* [Swiss Society of Natural Sciences] purchased the *Grosse Fluh* in Steinhof for the equivalent of CHF 2,500 to protect it against destruction.[21] This was followed by an "erratic-boulder fever" amongst the scientific elites to protect and study these rocks, which laid the foundations for the Swiss nature-conservation movement.[22] The impressive and "endangered" rocks also sparked romantic sentiments in the wider population, and they became popular tourist destinations or memorial stones for poets and other important public figures.[23] For less iconic erratics, the opening of the Gotthard train in 1882 brought some respite to the widespread "harvesting" of erratic boulders, as hard rocks could now be mined in the Canton of Ticino and along the rail tracks and easily transported by rail.[24] Nonetheless, removal for building materials and field clearings continued, so that by 1909 the Swiss Society of Natural Sciences was forced to also purchase the *Kilchliflüeli* rock in Steinhof after reports of a continuous removal of erratic boulders from surrounding fields.[25] At other sites, such as the forests near Feldbrunnen St. Nikolaus, the Canton of Solothurn placed more than 200 blocks under protection in 1893, providing them with metal plaques. In 1917, a survey of erratic boulders in the same canton followed, documenting a further 500 erratic boulders.[26] By the mid-twentieth century the protection of erratic boulders was well established as an important pillar of Swiss national heritage and nature conservation.[27] Nonetheless, numerous rocks still bear the marks of their attempted destruction in the form of boreholes and split-off surfaces. As vegetation and erosion make these marks increasingly indistinguishable from

A boundary stone in front of an erratic boulder at Steinhof. Originally, the erratic boulder marked the border; in 1764, a boundary stone was added to mark the border between the Canton of Solothurn and the Canton of Bern. Photo © VOGT Landscape Architects

Erratic boulder as a memorial for Swiss geologist Amanz Gressly (July 17, 1814 – April 13, 1865) in the Verena Gorge in Feldbrunnen St. Nikolaus. Photo © VOGT Landscape Architects

those of Neolithic or Roman times, they remain symbols of how humans are constantly altering and redefining their physical and cultural landscapes.

A Long Way from Home

Clinging to the edge of a rock outcrop, we find grey-cushioned grimmia (*Grimmia pulvinate*) and Hedwig's moss (*Hedwigia ciliata*). These delicate lichen species have found their way onto the *Grosse Fluh* through spores that travelled hundreds of kilometers from the Southern and Central Alps to the Swiss Central Plateau (*Mittelland*). More than 200 moss and lichen species have come to grow on these imposing boulders, which act as habitat islands widely scattered across the landscape.[28] Most erratic blocks are chalk-free, made up of granite or gneiss and, therefore, offer the same geological conditions as the High Alps in which this flora is usually found.[29] Each boulder hosts a unique species assemblage, influenced by the geological composition of the rock material, but also the light conditions around the erratic boulder, varying from dark forests to sunny, weather-exposed, open fields. The colonization of an erratic block takes hundreds of years, and each population of species remains isolated with no genetic exchange with other rocks.[30] Yet the flora, and the rocks also share a more recent common history: in the nineteenth century, when Swiss scientists debated the newly developed theory of the ice age, they viewed the flora as further proof that large masses of ice had moved these rocks from the Central Alps into the Swiss Central Plateau.[31] During the "erratic-boulder fever" that ensued amongst the educated elites, many people took part in field trips to inspect erratic boulders and collect their flora, driving some of the species close to extinction. While still common in their original habitat in the Central and Southern Alps, the flora found on erratic boulders in lower parts of Switzerland is critically endangered today. Bouldering enthusiasts "clean" the erratic blocks of their flora to have better climbing conditions, and the growth of shrubs and trees around neglected rocks shade out sun-loving lichen and mosses.[32] Increasing air pollution and the removal of erratic boulders from agricultural fields are further threats.[33]

1 Steven Mithen, *After the Ice: a global human history, 20,000–5,000 BC*, Harvard University Press, Cambridge MA, 2004.
2 Walter Moser, "Findlinge im Kanton Solothurn, Zeugen zweier Eiszeiten," *Jahrbuch für solothurnische Geschichte* 67, 1994.
3 See note 2 above.
4 Urs Schwegler, *Was sind Schalensteine?*, SSDI Schweizerisches Steindenkmäler-Inventar, 2016.
5 Stephan Pinösch, "Allgemeine Natur der Schalensteine," *Jahrbuch für solothurnische Geschichte* 14, 1941.
6 Karl Ludwig Schmalz, "Steinhof – Steinenberg," *Jahrbuch des Oberaargaus* 9, 1966.
7 See note 4 above.
8 See note 5 above.
9 See note 6 above.
10 See note 6 above.
11 Juri Jaquemet, "Kulturspuren auf dem Jolimont," *Seebutz*, 61, 2011.
12 Dieter Schneider, Erich Gubler and Adrian Wiget, "Meilensteine der Geschichte und Entwicklung der Schweizerischen Landesvermessung," *Geomatik Schweiz* 11, 2015.
13 See note 6 above.
14 See note 6 above.
15 AWA Amt für Wasser und Abfall, Abteilung Gewässerregulierung, *Regulierung der Jurarandseen – Grundlagen und Vorgehen*, Bern, 2012.
16 See note 2 above.
17 Emanuel Maurer, "Im Interesse der Wissenschaft und zur Ehre des Landes: der Schutz der Findlinge im Kanton Bern," *Mitteilungen der Naturforschenden Gesellschaft in Bern* 62, 2005.
18 See note 17 above.
19 Evelyn Boesch Trüeb, "Geologie auf der Wiese: der Luegibodenblock von Habkern," *ETHeritage*, 2020.
20 See note 2 above.
21 See note 6 above.
22 See note 17 above.
23 See note 19 above.
24 See note 2 above.
25 See note 6 above.
26 See note 2 above.
27 See note 17 above.
28 Daniel Hepenstrick, "Findlinge sind wertvolle Lebensräume," *Umwelt Aargau* 84, 2020.
29 Raymond Delarze and Yves Gonseth, *Lebensräume der Schweiz*, Ott Verlag, Thun, 2008.
30 See note 28 above.
31 Daniel Hepenstrick, *Projektbeschrieb: Naturschutzbiologie der Findlingsflora*, Zürcher Hochschule für Angewandte Wissenschaften, 2019.
32 See note 28 above.
33 See note 31 above.

Field Trips –
Learning from Landscapes

Maren Brakebusch

In our work, field trips represent an important approach
to new topics and questions. Here, searching and researching
describe both the method and the intention. For if you
study the landscapes through which you walk and hike, your
initially distanced gaze changes into an intensive contem-
plation, a literally physical perception and, ultimately,
results in a precise description of such landscapes. In fact,
the very perspective of a walker or hiker is crucial, for the
motion this entails is not only a mode of experiencing space
but also a possible means to understand the landscape
that surrounds us. A field trip approaches the sequential
nature of a landscape on different levels. The very form
of our locomotion, the chosen speed or the distance covered
determine the direction of our gaze. Being accompanied
by someone else may well focus our gaze and knowledge
additionally due to a shared discourse about the landscape.
Our knowledge of a place and our approach to a landscape
cause us to become aware of its specific aspects at
a certain point in time. Thus, "distance and engagement"
describe not only our musings about design work in the
office but also a methodical approach to this way of experiencing a real landscape on site.

Our annual office excursion is designed as a field trip (search)
without any specific intention. It serves no immediate
purpose, but takes place out of a general interest in a land-

scape or specific phenomena. The search is preceded by an intensive preparation, based on the size of the group and the desire for an instant and intense experience. Unplanned happenings play an important role in the highly literal sense of – *what happens to you by mere chance.* On our field trip to the Val-de-Travers and the Glacière de Monlési, the largest ice grotto in the Neuchâtel Jura and, thus, the most powerful glacier in Switzerland outside the Alpine region, the contrast of meter-thick ice next to lush, fern-covered rock faces left us with the unexpected realization that, on our way there, we had found a *landscape compass* on the trunks of the surrounding forest. The moss-covered side of the tree trunks showed us the way to the north, while the lichens pointed to the west, protectively turned away from the prevailing east wind in the Jura mountains.

A specific field trip that is primarily carried out within the scope of an actual project with the intention of investigating and revealing the manifold aspects of a location is based on a detailed study of the general conditions of the field trip plus the prior acquisition of specific knowledge. The actual field trip (research) then focuses on a targeted checking, measuring, or matching on site, with the knowledge gained from this essentially determining the contents and expression of the design work. Confronted with the task of creating a forest with different atmospheres on as small a space as possible, our systematic surveys of four forest scenes we found on our field trip to places near Basel helped us to develop a vocabulary for the design. The varying distances between the tree trunks create equally differentiated densities and configurations in space that determine light and shadow. Translated into diagrammatic models, photographed, measured, and categorized, their situational use and partial superposition resulted in specific forest situations for the Novartis Campus park in Basel.

However, what is measurable does not always reflect the actual, real experience. In fact, the unexpected and often immeasurable must also be "captured." The localization and determination of a phenomenon is followed by the recording of the experience – an **isolation** – and its detachment from the context. This operation allows us to transpose the model-like quality of a landscape. To record the real landscape experience, the *travelling (landscape) architect* has recourse to a highly varied repertoire of tools. Their characteristic choice needs to be well considered before-hand regarding the possible result but also concerning the type of of locomotion. To digitally capture the *colors of a landscape* for a later visualization of the project, all that may be needed is a quickly downloaded app; in contrast, the watercolor to be colored calls for physically taking along pieces of rocks and plants that embody the tonality of the site and transmit it to the design.

The model-like nature of our landscape designs requires an **abstracting** of the extracted sample of the landscape referred to. Thus, the cut herbaria that, with their 1.60 meters, are almost as high as humans, are the dried result of samples drawn from real meadows during our research phase, which were further simplified by extracting certain

of their elements. The impressive result serves to successfully communicate the idea, but in order for it to be a true
model for the landscape design, requires a comparison and
a condensation on the basis of our knowledge of a year-
round image, which meets the desired effort of maintenance
and intensity of use.

Landscape as a model does not mean a reduction of scale
in the proper sense, such as generally precedes architectural
models. Rather, the goal of the **translation** is the creation
of several "approaches." An actual experience of landscape
helps us to read it. The moss-covered, large-scale tuff
blocks in the courtyard of the Zurich Hyatt has us perceive
a forest without whose protective canopy the moss could
hardly grow in such an exposed place. But even lacking this
knowledge, we may still enjoy what we see in front of us.
On an aesthetic level, those who only see the moss enjoy its
soft texture in contrast to the precisely cut tuff blocks.

To walk through a landscape requires, besides the actual
execution, a purposeful preparation as well as the
translation step in the post-processing. In addition to all
the planning, you need to remain open for the unexpected
in the expected.

Paris Agreement

2010

Rolling Stones

Günther Vogt, Violeta Burckhardt, and Simon Kroll (VOGT Landscape Architects) as part of the *Future Assembly* exhibition by Olafur Eliasson's and Sebastian Behmann's Studio Other Spaces at the Central Pavilion, Giardini della Biennale di Venezia

Photos by Alessandra Chemollo

In landscapes formed by glaciers, erratic boulders bear witness to the former presence of ice and, at the same time, refer to the distant places of their actual origin. However, the respective knowledge is pretty recent. Until the middle of the eighteenth century, people marveled in disbelief at the very presence of these boulders whose material and dimensions differed quite noticeably from their surroundings. Some quite adventurous stories were told, and explanations "found" in an attempt to shed light on this mystery.

The same as the erratic boulders of the various glaciers, the stones of the exhibition come from the most diverse Alpine formations. They reflect the range of geological conditions of this Alpine space, whose origins go back to the collision and partial overthrust of the two continents of primeval Africa and primeval Europe. Geomorphological processes subsequently created the characteristic relief and the specific and fundamental conditions for the genesis of Alpine culture.

This contribution to the exhibition allows you to directly and physically perceive the movement of the stones with your senses. You can sit on them, touch them with your hands, feel their texture and temperature, or shift them almost effortlessly with a simple motion. No longer through the flow of the ice, but by the sheer force of man, these giant boulders weighing virtually tons thus advance in a horizontal direction. Bringing them to a standstill requires a repeated effort though.

Visitors change their roles from observers to actors: emblematic of the species of *Homo sapiens*, which – by now – changes our landscapes on the same scale as the forces of nature did. The lonely man in the center of Julian Charrière's picture *The Blue Fossil Entropic Series Pole* (shown as part of the contribution to *Common Water – The Alps*), who tries to melt the iceberg with his flamethrower, illustrates this very fact. At first glance, his plan seems as hopeless as that of Sisyphus, the tragic figure from Greek mythology. As punishment for his deeds, Sisyphus had to roll a boulder up a mountain in the underworld, which rolled back down into the valley every time he reached the top. However, if one shifts the perspective from an individual human to the almost eight billion people on planet Earth, global consequences will inevitably impact our world.

The traces of our actions are deposited as sediments on Earth, and recently so. Humans are thus identified beyond any doubt as a geological force. Uncovered by later generations, these layers will be interpreted as part of the history of human existence and provide information about our relationship to Earth's landscapes, characterized by a one-sided use of its resources.

Future Asse

The Medial-Moraine Model – Complete Revision of a Century-Old Ideology

Autobiographical Notes by Gerhart Wagner

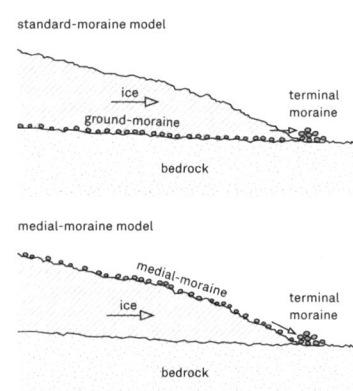

standard-moraine model

medial-moraine model

Standard-moraine model: The rock material stems from the bedrock and is pushed along underneath the glacier. If the end of the glacier remains the same (during a longer-lasting "stadium"), it is pressed together as a terminal moraine. (By the way: the glacier always flows, even if its end "stands still!") Medial-moraine model: The rock material stems from the rocks above the glacier and is taken along on the surface of the ice. If the end of the glacier remains constant, it accumulates as a terminal moraine. Schematic diagrams © Gerhart Wagner, 2021

From this they deduced the truest law,
turn lowest into highest, to be sure.

Mephistopheles, in: J. W. Goethe, *Faust,* Part II, Act IV

Preliminary Note

My academic training was actually as a biologist. Therefore, I am often asked how I came to be involved in geology and to consider a term to be so important that had hardly ever appeared in the literature on the ice age. To understand this, some insight into my curriculum is needed. Thus, I will try and present the relevant facts from my curriculum vitae to show how it fell to me to clarify an unsolved question of glacial morphology in the area I call home, by taking up the paradigm of the *medial moraine*, and how this term subsequently proved helpful for many other fields as well. But I shall also show how the model, after an initial recognition by expert circles, met with bitter resistance.

Autobiographical and Historical Roots

Childhood and School Years 1920–1939

My parents' house was in Bolligen near Bern, a scant hour northeast of the city. To the south, we could see the Alpine ridge with the Stockhorn, and to the east, barely a kilometer away, a hill with a round brow and the funny name of Hühnerbühl or Chickens' Hill. On its slope, you could recognize a gravel pit, which belonged to Niklaus Kunz, a farmer, from whom my father got the gravel he

used for the access road to our house. On the round knoll, which we sometimes climbed on a Sunday excursion, there was a strange cement pedestal of about two square meters, right in the middle of a field: here, it was said, an astronomical telescope had been placed before the World War, but that Peter Stuker, the astronomer who had used it for his purposes, had moved away. Along with him, the apparatus had also disappeared.

Also within the radius of our excursions, was the Stockeren Forest with its large blocks of stone that our teacher called *erratics*, whatever that name was supposed to mean. He also spoke of the *Ice Age,* the glacial period. But what were we to imagine it was? Later, I attended grammar-school in Bern. The subject of natural history also taught us about geology. The teacher in charge, one Dr. Steiner-Baltzer, did this in the form of an appealing local scientific history. The terms *erratic boulders* and *ice age* were now allocated a specific meaning, and an important other term was added as well: the *moraine.* We were told about lateral moraines, terminal moraines, and ground moraines covering a large area.

At the hill called Längenberg south of Bern, there are prime examples of moraines, which we visited on excursions. On the wall of our classroom hung a map of Bern and its environs with the moraines drawn in, in particular the arc of moraine hills circling the city center. These moraines were plausibly explained to be the remains of a once continuous end-moraine wall of the Aare Glacier in the so-called Bern-Stadium. Local geologist Dr. Eduard Gerber, who was still alive at that time, stated that five

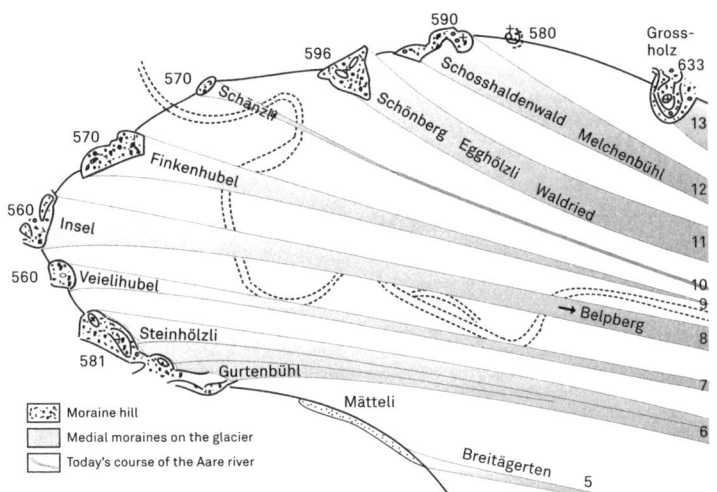

The moraine hills around the city of Bern: are they remnants of a once continuous end moraine or end aggradations of individual medial moraines? The illustration shows the end of the Aare Glacier snout in the Bern Stadium. Wagner, *Mittelmoränen*, 2014, p. 75.

more such stadia could be recognized in the Aare Valley. Unfortunately, the moraine conditions in my residential area of Bolligen remained somewhat mysterious.

Such insights into the formation of the landscape of our immediate neighborhood were highly impressive but were only a small facet of our grammar-school lessons for a Matura Type A education (college preparatory school) with Latin and Greek as main subjects. For now, completely different things became suddenly important: on our very last day of school, on September 1, 1939, Hitler invaded Poland. That meant war – probably for quite some time – and the Federal Council ordered a general mobilization of the Swiss Army.

Studies 1939–1949

After graduating from grammar school, I decided to become a teacher because I considered teaching my best talent. Therefore, I first of all considered the then Teachers' College, intending to train as a secondary-school teacher. There were two professional directions: given my Matura graduation, opting for a linguistic-historical degree would have been logical, but I chose a mathematical-scientific track instead. However, in the years to come, military service became mandatory. In addition to the obligatory instruction services, there were also months of active service due to the ongoing war. Fortunately, these often took me to places of scenic and geological interest. It was a great feeling when I was able to pass the secondary-

Upper and Lower Grindelwald Glacier.
Colored aquatint, ca. 1850, ETH Zurich,
ETH Library, Collections and Archives /
Image Archive

school teacher's exam during a vacation in the fall of 1943, while still on active duty. For further studies, intending to become a grammar-school teacher, I chose Zoology and Botany as my majors. However, as mentioned above, geology was also taught as part of natural history at the Bernese grammar schools, so I took the precaution of choosing Geology as an additional minor. I attended – as far as my active service allowed – the corresponding lectures and excursions, but did not sit any exams in Geology. In 1946, I obtained a grammar-school teacher's license for the subjects of Zoology and Botany with a minor in Physics and, in 1949, obtained a PhD with a dissertation in Zoology.

Career 1949–1983
*Secondary-School Teacher in Grindelwald 1949–1950 and
Grammar-School Teacher in Bern 1950–1958*
It was a stroke of luck that I got a job as a secondary-school teacher in the glacier village of Grindelwald in the Bernese Oberland. The two Grindelwald glaciers, the Upper and the Lower Glacier, were still easily accessible at that time and duly studied by me. Here, ice-age conditions were demonstrated in every detail to me *ad oculos*.

Only a year later, the position of my former teacher of natural history at the Kirchenfeld Grammar-School in Bern became vacant and I was appointed to be his successor. I now had to teach Geology as well as Biology, which I did with pleasure. My additional studies now came in handy and I thoroughly familiarized myself with the subject matter of geology. As a teacher, I could now use the illustrations I had already known as a student. During my eight years as a teacher at the Kirchenfeld Grammar-School, geology became as dear to me as biology.

In 1958–1964, I was a Federal Official for Radiation Protection; and in 1964–1969, Assistant Professor of Zoology at the University of Zurich. During these 11 years, I was far away from school and moraines. But soon, they were to once again become an important topic for me!

Principal of the Bern-Neufeld Realgymnasium 1969–1983
In 1969, I became Principal of the meanwhile newly established Neufeld Realgymnasium, a Swiss secondary,

university-preparatory school in Bern emphasizing science, maths, and modern languages. I held this position until I retired in 1983. During these 14 years, I had a compulsory teaching load in addition to my work as Principal. Once again, my subject was Natural History and, consequently, geology once again became a topic, too. Only, I now tried to take a closer look at the delicate moraine conditions in my old home area near Bolligen. Eduard Gerber, the local geologist already mentioned above, had published a study about this area in 1955, one year before his death. He found that the Aare Glacier was, at a certain stage towards the end of the last ice age in the area of Bern, deflected by Rhone ice and had to have entered the Worblen Valley with its ice tongue from below. This was quite plausible. But Eduard Gerber did not get to finish his line of thought. It fell to me to do so: I told myself that the Aare Glacier had to have formed a medial moraine with the Rhone Glacier, and that this might be detected in the field. But where should I look for it? I pondered the question, looking at the large wall map in my Principal's Office. It had to be in the area of Bolligen, perhaps a bit further south, in the area of the Hühnerbühl hill, and eureka! It *was* the Hühnerbühl!

I was delighted by this lightning-fast conclusion, subsequently thought it through as to all its aspects and found it to be watertight, though I would never have arrived at it without the last publication by Eduard Gerber. The discovery was first made public by a newspaper article in the "Bund." It was not disputed either, and I made guided tours in the field with geologists and geographers, and with anybody else interested. The paradigm of the medial moraine, which had been completely missing until then, was now also included in the explanations to the geological map sheets of Worb and Bern (both of 1999). I even received a letter of thanks from the geologist responsible for mapping the Worb sheet.

Retired Since 1983
First of All Botany
After the early termination of my professional career, my botanical work with biochemist Konrad Lauber began. The publication of the *Verbreitungsatlas der Farn- und Blütenpflanzen der Schweiz* by Max Welten and

Ruben Sutter (1982) provided the incentive for our plan to create an illustrated Flora of the Bernese Oberland. I had sorely missed having one during my botany studies. Since I had collected plants wherever I went and established a quite important and extensive herbarium in the four decades that had passed since, I felt able to embark on this project. Konrad Lauber, for his part, was one of the best plant photographers in the country at that time. With his Nikon camera and Kodachrome 64 films, he had been photographing plants for a good three decades and had already produced thousands of top-quality photos. We went to work and took our time. What was planned as a Flora of the Bernese Oberland became a Flora of the Canton of Bern in 1991 and *Flora Helvetica* in 1996. Konrad Lauber managed to get all 3,000 species of the *Flora Helvetica* except one in front of his camera. And I wrote the texts. For geology however, I had little time during these years. At least, I once again looked around in my old home area, the Worblental or Worblen Valley, where I had resettled in 1981, and discovered more medial-moraine structures. You can use your very same eyes, after all, to look at flowers

Sheet from the herbarium of Gerhart Wagner © Gerhart Wagner, 2021

At the sight of this enormous moraine hill, the "scales fell from his eyes," thus Prof. Dr. René Hantke. The picture shows the Wilerhubel, a 1-km-long and 70-to-100-meters-high moraine hill. The arrow points to the impact direction of the medial moraine. Wagner, *Mittelmoränen,* 2014, p. 81.

or moraines. But only after the completion of the *Flora Helvetica*, I was once again able to focus on geology.

A New Attempt at Geology –
Prof. Dr. René Hantke Appears on the Scene

The paradigm of the medial moraine proved to be unexpectedly productive. I had published the first findings on the Hühnerbühl hill in 1986 in the "Mitteilungen der Naturforschenden Gesellschaft in Bern." After I had searched the rest of the Bernese Plateau for medial moraines, I ventured a second publication in the Bernese Mitteilungen in 1997. I also sent it to René Hantke. He had found my first paper of 1986 plausible and already then visited the Worblental (the valley in which the Hühnerbühl hill is located) to have, guided by me, a close look at it. Now he visited my home area a second time and I guided him to the moraines at the hill called Längenberg. At the sight of the Wilerhubel, the enormous moraine hill near Riggisberg, he later averred, "the scales fell from my eyes." He now recognized the far-reaching significance of the medial-moraine paradigm and became my staunchest fellow campaigner. Medial moraines, he said, were exactly the element that had been lacking when he wrote his three-volume work *Ice Ages* (1978–1983).

In Search of Additional Medial Moraines – Exemplary Situations

I now dared to take a closer look at more remote areas of our country in the area of ice-age glaciation. I did not have to search for any existing glacial structures as I found them on the existing geological maps of the areas to be studied. I borrowed them from the Swiss Geological Survey

Moraines in the Aletsch Forest: Here, the left medial moraine of the Great Aletsch Glacier overflowed during the "Little Ice Age." The arrows show the impact direction of the medial moraine. Wagner, *Mittelmoränen,* 2014, p. 103.

in Bern. I was now able to specifically locate the moraines in the field and study their morphology in situ in order to assess them and, if necessary, identify them as medial moraines. Now I also dared, and successfully so, to offer my publications on the subject of medial moraines to the international scientific organs "Eclogae" (2001) and "Zeitschrift für Geomorphologie" (2003) (see Publications).

Medial Moraines in Switzerland
Medial Moraines in the Canton of Valais
The Canton of Valais, which was particularly familiar to me from my active military service, was my next target. In the glacial monuments of Massegga (a monstrous hill near Naters), which projects far into the Rhone Valley, I found a first and splendid example. Can medial moraines really be so large? I often had to remind myself: Dare to think! And yes, they can. It was easy to identify numerous other cases of medial moraines of all sizes in the Rhone Valley and in the side valleys, from small hills to large landscape-sculpting formations.

Medial Moraines in the Canton of Zurich
I now turned to the Canton of Zurich. A particularly interesting case were the moraine hills in the very city of Zurich, which are now built over or eroded. As with the hills around the city of Bern, the acknowledged doctrine in Zurich is that they were remnants of a once continuous terminal-moraine snout or edge, which once upon a time dammed Lake Zurich to a 10 meters higher level than today. At least, this is what the Zurich textbooks say. In the light of the medial-moraine model, these hills, as in Bern, are clearly the terminal accumulations of several medial moraines. They never dammed Lake Zurich though.

In the southern part of the Canton of Zurich I found a large number of, to me, quite obvious medial moraines from the Linth and the Lake Walen arm of the Rhine Glacier. But there I unfortunately came into conflict with Thomas Gubler – and there my suffering began. But of this later.

Medial Moraines in the Cantons of Vaud and Geneva
On the Swiss side of Lake Geneva, as well as in the area of the Canton of Geneva and on the lake bottom, I identified

a large number of glacial landscape structures as medial moraines, from small individual structures to imposing landscape structures of more than 20 km length. By chance, I stumbled upon a publication by Adrien Jayet (1966) entitled *Résumé de Géologie glaciaire regionale*. Jayet had studied the glacial formations around Lake Geneva for decades and was quite unhappy about the fact that all glacial formations that could not be explained as lateral moraines were generally declared to be ground moraines by the current school of thought. He looked for various other possible interpretations and also considered medial moraines.

A Medial Moraine in the Canton of Ticino

The mountain ridge running from Monte San Salvatore to Morcote with its highest points at Cima Pescia (835.4 m a.s.l.) and Monte Arbostora (822.3 m a.s.l.) has a two-km-long moraine structure in its southern part. Here, a medial moraine has been pushed over an island protruding from the ice (in the Inuit language: on a *nunatak*). Its axis strikes into emptiness in a northeasterly direction towards the northeastern arm of Lake Lugano, indicating that it is a medial moraine of the Ceresio arm of the Adda Glacier. This large structure superimposed on a mountain is completely inconceivable as a ground moraine.

The Rhine Glacier with its presumed medial moraines at the Constance Stadium. Wagner, *Mittelmoränen*, 2014, p. 124.

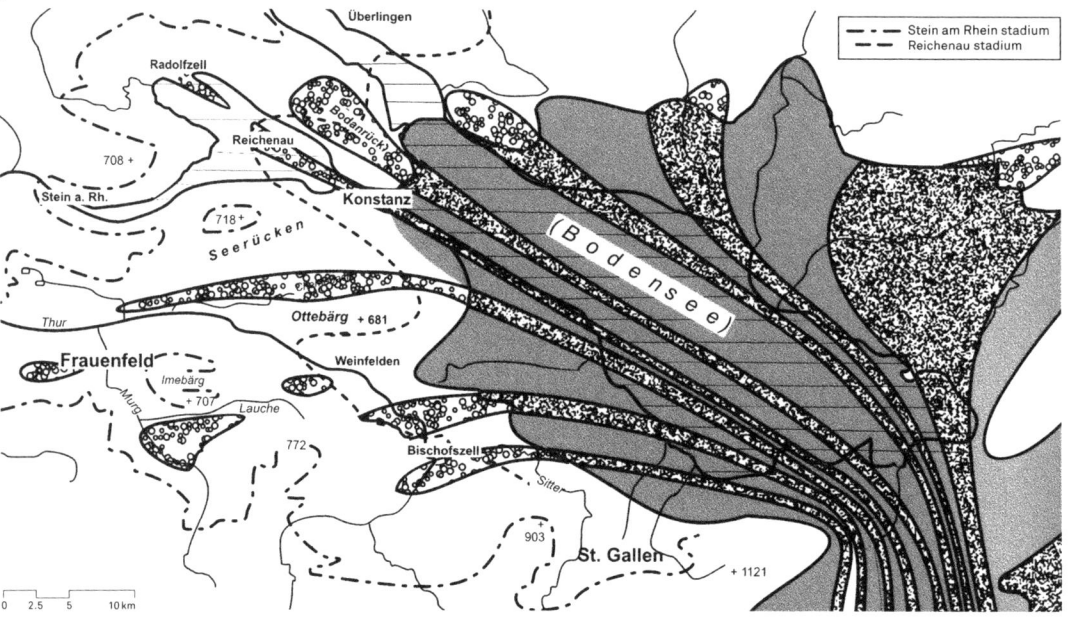

A Medial Moraine in Germany

Reichenau Island in Lower Lake Constance is mapped on the geological map of the District of Constance as Würmian "gravel deposited near the edge of the ice, alternating with or covered by moraine material," with the two highest elevations, Hochwacht and Vögelisberg, about 40 m above the level of the lake, as "drumlins in a ground moraine." To understand such a thing is difficult. The morphology of the island, this enormous ice-age monument, shows the characteristics of a large medial-moraine accumulation at its snout: "spurs" rising in the flow direction, two culmination points, a (double) concave front with "fingers" extended in the flow direction.

Medial Moraines in Italy

Near Ivrea in the Aosta Valley in the Western Italian Alps, are some of the most powerful moraines in Europe. Naturally, this area also has some important medial moraines. The overall picture of the huge moraine complex of Ivrea shows several inwardly protruding medial-moraine spurs. Penck and Brückner state: "The flanks of the frontal moraine arcs, which are joined to each other, jump from the southeast into the wide accumulation plane *like medial moraines*" (Albrecht Penck and Eduard Brückner, *Die Alpen im Eiszeitalter,* 1909). This is one of the very few places where Penck and Brückner use the term of "medial moraine." But even here, a clear interpretation as medial moraine is not considered, though it literally imposes itself.

The Controversy of the Medial-Moraine Model

Introduction to Geological Circles

After 1998, René Hantke introduced me to Quaternary geological circles. On September 7, 2000, I was given the opportunity to present a short paper at a meeting of DEUQUA (German Quaternary Association) in Bern. I tried to plausibly present the medial-moraine concept in 20 minutes to Quaternary geologists from Germany, Austria, and Switzerland and got a lot of applause. The then president of DEUQUA, Prof. Dr. Christian Schlüchter from Bern, also allowed me to come along on the excursion to Valais the following day, where I was able to point out some beautiful medial moraines in the field.

Meanwhile, I had also became acquainted with Prof. Dr. Karl Albert Habbe, President of AGAQ – the German Arbeitsgruppe Alpenvorland-Quartär or Work Group Quaternary Alpine Foreland – and René Hantke took me to events in Germany, once to Ottobeuren, where I was allowed to give a short lecture and join an excursion. Karl Albert Habbe was convinced of the validity of the medial-moraine approach, which he had already come up with himself, and showed me a beautiful example in his own area. He told me about the Lampolding drumlin field (part of the municipality of Kirchanschöring in the Upper Bavarian district of Traunstein), which had quite certainly been carved out of the medial moraine of the Saalach and Salzach Glaciers.

At that time, the medial-moraine model seemed to be a triumph in many ways. Even Christian Schlüchter, the expert I considered most important for my arguments, a full professor of Quaternary Geology at the University of Bern, did not question its plausibility. But the punishment for a biologist who ventures into geological realms and doubts generally accepted views was not long in coming: from Zurich, a damning report reached me.

The Conspiracy Against the Medial-Moraine Model and Problems with Christian Schlüchter

Ever since the time of his doctoral thesis, I had had a collegial relationship with Christian Schlüchter. I had often drawn his attention to new terrain outcrops, which we had a look at together. The break with him occurred in connection with my 2002 paper published by the Zurich "Vierteljahrsschrift der Naturforschenden Gesellschaft," to which I had submitted a manuscript with the title of "Eiszeitliche Mittelmoränen im Kanton Zürich, 1. Teil." The two editors of this renowned Zurich journal were Prof. Dr. Conradin A. Burga and Prof. em. Dr. Frank Klötzli. Conradin A. Burga knew that my work was something completely new and had advised me in a letter to "consult a few more experts on the subject" before publication and gave me the names of Dr. Max Maisch and Dr. Georg

Wyssling. I then sent my publication to these two gentlemen and also informed them that I worked on a similar publication about the Canton of Zurich.

In December 2002, when Part I of my Zurich paper was already in print, Thomas Gubler, a geologist, whose name I had not encountered in either the older or the newer Zurich Quaternary literature, turned up quite unexpectedly. He had graduated in 1989 with a Molasse thesis under René Hantke and, in the meantime, had become an important man, working on the map sheet of the Albis on behalf of the Swiss Geological Survey. It was a pity that I had not known this earlier, otherwise I would have contacted him. He was now understandably angry about me as an intruder in "his" area (the Albis map sheet appears extensively in my Zurich work). For the attention of Conradin A. Burga, he had thereupon written a scathing critique of my Zurich manuscript. But it was too late to change anything in my text.

The editors of the Zürcher Vierteljahrsschrift must have been shocked by this scathing review of my work. Shortly before, Frank Klötzli had expressed his enthusiasm about my work on medial moraines and congratulated me on it. He even coined the expression "medial-moraine model" for our approach. René Hantke and I took over this designation and, therefore, called the previous opinion the "standard model."

Now this low blow by Thomas Gubler. I received his damning review, co-signed by Georg Wyssling, on January 6, 2003, with a letter by Conradin A. Burga. He asked me to make "the necessary relativizations or corrections" for Part II and announced a review of Part II by Mr. Gubler and Mr. Wyssling. Thomas Gubler's critique, however, hardly mentioned anything that could squash my interpretations. I considered his most substantial objection, that "in the last glaciation demonstrably no Linth ice came into the Knonauer Amt," (thus, a medial-moraine formation from the confluence between the Linth and Reuss Glaciers could be excluded), to be demonstrably wrong on the basis of the already existing literature and given my own observations in this regard. I wrote a corresponding

The Reuss, Limmat, and Thur Glaciers with their presumed medial moraines in the Zurich Stadium. Dr = locations of present-day drumlin fields. Medial moraines 1–11 on the Linth Glacier are, according to their end aggradations, as follows: 1 Hirzel, 2 St. Anna-Hügel, 3 Lindenhof, 4 Hohe Promenade, 5 Itschnach, 6 Wetzwil, 7 Toggwil, 8 Buechholz, 9 Forch, 10 main strand of the Glatt Valley, 11 Russikon. Wagner, *Mittelmoränen*, 2014, p. 123.

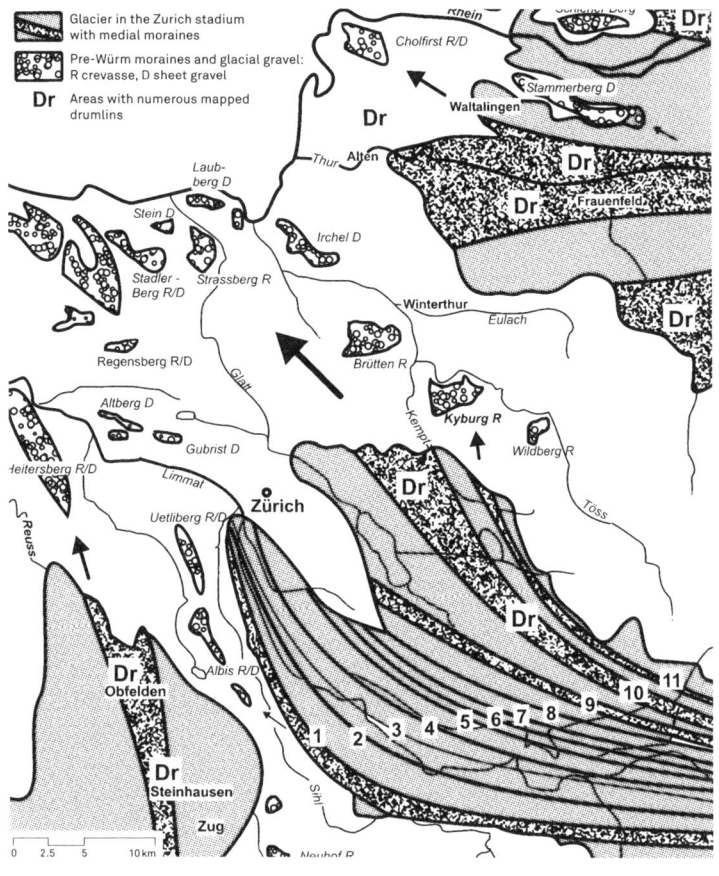

replica to Thomas Gubler with copies to Zurich and Christian Schlüchter. To the latter, I suggested to make the Knonauer Amt a test case for the medial-moraine model.

The Excursion of April 9, 2003
While this low blow was already a completely new experience for me, it turned out to be only the beginning of a descent into true purgatory. After April 1, Thomas Gubler called to arrange a walk-through of the Knonauer Amt together with Christian Schlüchter and René Hantke. I thought this was a good idea as it also matched my wishes, and I imagined, I would be able to explain the medial-moraine model to the gentlemen on the Hirzel in view of the conditions there, which seemed so clear to me.
So we drove in Thomas Gubler's car to the points he had selected. He showed us blocks of Gruon Valley conglomerate, a bedrock of the Reuss Glacier, at various places. Thus, he proved credibly that the Reuss Glacier had been there. But did he also prove the absence of Linth Glacier

ice? At the fourth site, a heated discussion ensued between Christian Schlüchter and me, in which he accused me of being unscientific and cast doubt on the term "secondary lateral moraine" used by me (moraine walls formed by the emergence of medial-moraine sections at the edge of the glacier). I was not able to play my cards at all and René Hantke was not really "there" either. Thomas Gubler and Christian Schlüchter had us in something of a "choking hold." And the worst was yet to come: Thomas Gubler now led us to the hill between Ebertswil and Husertal and showed us that it consists of Molasse and not of moraine, as I had stated in my Zurich paper, based on the geological map of the Canton of Zurich of 1968. This was definitely the worst blow of the day. I felt truly devastated and even called it a "total defeat" at the final meeting.

Recovery from the "Total Defeat"
Should all my work on the medial moraine really count for nothing? I did not think that was possible, because I already had too much solid material. After a few days of great contrition, I first tackled the biggest embarrassment, the question of the Molasse ridge or a moraine between Ebertswil and Husertal. Had there really been no moraine? On closer inspection, I found that the geological maps of the area show a pit above the village of Ebertswil. So I went there on Easter Saturday, April 19, and saw quite clearly that a pretty large former pit, today a biotope, is embedded in a moraine. Unfortunately, the wall is no longer open, but some of the boulders lying around are Verrucano, the bedrock of the Linth Glacier. So there were two findings in my favor. Firstly, the moraine is no doubt there, but its axis does not run over the ridge, but about 250 m further southwest. And secondly, Linth erratics are present. These two findings helped me to get over the moral and emotional rollercoaster. From then on, my mood definitely improved.

But for the time being, I only received a memo from Thomas Gubler about the field inspection, which recorded my "complete defeat" at that time. I could not accept that. But before I proceeded to comment on it, I once again went to the Knonauer Amt on another search for Verrucano *in situ*. This time I chose Jungalbis (a place in the southern part of the Albis chain) whose southeast-facing, wooded spur had, in my opinion, to have been brought there by the Linth Glacier. After a long search in the forest on this spur, I found a Verrucano block with a visible volume of of $80 \times 40 \times 20$ cm. It was quite exhilarating to recognize this block in the darkness of the forest, and I greeted it like a good old friend. I took the liberty to cut a hand-sized piece from it with my geologist's hammer and, later, had it confirmed by René Hantke as Verrucano! So Thomas Gubler's claim that, during the last glaciation, no Linth ice flowed into the Knonauer Amt is demonstrably erroneous.

It was only now that I wrote and submitted my statement on Gubler's memo. I admitted having made a mistake concerning the moraine near Ebertswil, which I would correct in my next publication (Zurich Part II). But I firmly rejected Thomas Gubler's thesis concerning the absence of Linth ice in the Knonauer Amt. I also sent the statement to the Zurich editorial office.

During this time, two other manuscripts on medial moraines were already on their way: the second part of my Zurich study and an extensive manuscript by Hantke/Wagner on medial moraines in the Canton of Thurgau, the latter for a special volume of the Thurgauer Mitteilungen. In both papers, besides the drumlins, the mysterious "Höhenschotter" was taken into account in the medial-moraine model. From Zurich I received the gratifying information that my second part would also appear in full in the quarterly. The Thurgau manuscript had come under fire from the *Landesgeologie,* the Swiss Geological Survey: as they co-financed the special volume on Thurgau, they demanded changes to our text. As far as somewhat more cautious formulations were concerned, we agreed to accept them, but in no case would we accept cuts in the basic statements concerning the medial-moraine model. The paper was published in December 2003, but was relativized in the preface of the special volume with the formulation that it "contradicts previous views confirmed by new research results." I could live quite well with this editorial comment, after the much worse and scathing review from Zurich.

The complex of the Gorner Glacier (Valais, Switzerland) seen from the north as an example for a recent system consisting of glaciers of varying size and medial moraines. Only the two largest parts of the complex, the Grenz Glacier (Gr) and Zwilling Glacier (Zw), form the terminal tongue. The medial moraine between the Lower Theodul Glacier (Th) and the Triftji Glacier (Tr), as well as one medial moraine each **on** them are visible from start to end. The medial moraine between the Breithorn Glacier (Br) and the Schwärze Glacier (Sch) turns into a secondary lateral moraine and moves from the right-hand side of the Breithorn Glacier to the left-hand side of the Schwärze Glacier. The Gorner Glacier (Go) itself ends within the picture, while the medial moraine between it and the Grenz Glacier (Gr) turns into a secondary lateral moraine of the latter. The signum Y refers to two Y-formed stumps at the origins of medial moraines of the "Little Ice Age." Photo © J. Alean, Eglisau (Switzerland), 1983

Part II of the Zurich Publication and the Scathing Review by Graf et al. of 2003

In the September issue 2003/3 of the Zürcher Vierteljahrsschrift, the basically unchanged second part of my Zurich study appeared, but was immediately followed by an extensive article written by eight (!) partly influential authors (Graf, Burkhalter, Gubler, Keller, Maisch, Schindler, Schlüchter, and Wyssling) with the title "Das Mittelmoränenmodell in wissenschaftlicher Sicht." This declassified the medial-moraine model as something to be dismissed as unscientific. Actually, I was lectured about the nature of science. But the fact that the editors of the Vierteljahrsschrift (Conradin A. Burga and Frank Klötzli) had chosen a photo of the Gorner Glacier with its medial moraines supplied by me as the cover for the September issue, and considering their accompanying commentary, it was clear that they took the medial-moraine model seriously, in spite of the scathing review published in the same issue. In the publication of Graf et al., the medial-moraine model was effectively excoriated: absolutely nothing was considered to be correct about it – even the very beginning of everything, my interpretation of the Hühnerbühl hill as a medial moraine, which had not been doubted by anyone so far, was seemingly dismantled.

The arguments were astonishingly weak; it was not difficult to invalidate them and, in some cases, even reverse them. Medial moraines, so they averred, were merely features of active glaciers and disappeared along with the ice. That, in terms of material, they were much too weak to create permanent structures. The most important professional reproach was that I did not pay attention to the sedimentology of my medial moraines. This was true and I was fully aware of it. In my publications, I disregarded sedimentology, as it had been thoroughly researched by many authors, for it merely says something about the formation of a structure on the spot, but nothing at all about the all-important question of the long-distance transport of the material of which they are composed. In other words, the sedimentary character of a structure neither proves nor disproves the medial-moraine concept. Therefore, I can skip going into any more details.

Schematic compound ice-stream with numerous individual glaciers. Only the strongest form the snout of the ice stream, while smaller glaciers end on the margin of or between larger ones. Like each individual glacier, each medial moraine keeps its own character. All these moraines reach the ice margin, either at the snout of the glacier, or laterally, or at a nunatak (a rocky island in the ice), and deposit their debris there. A medial moraine can change into a secondary lateral moraine. Drawing by André Michel. Wagner, *Mittelmoränen*, 2014, p 16.

Ol local lateral glacier outwash of a medial moraine

Os overflow of a medial moraine that turns into a secondary lateral moraine

E terminal snout of smaller lateral and medial subglaciers

MA point of origin of Medial Moraines Type A: the confluence of their parental glaciers lies below the firn line.

MB point of origin of Medial Moraines Type B: the confluence of their parental glaciers lies above the firn line.

MN medial moraine terminates on a nunatak [island in the ice]

Mt terminal outwash of medial moraines

B ice-border section without surface debris

T track formed by the confluence of medial moraines

FD fluvioglacial deposits, formed from alluvial moraine debris

Lp primary lateral moraine

Ls secondary lateral moraine (overflowing medial moraine)

U confluent medial moraines

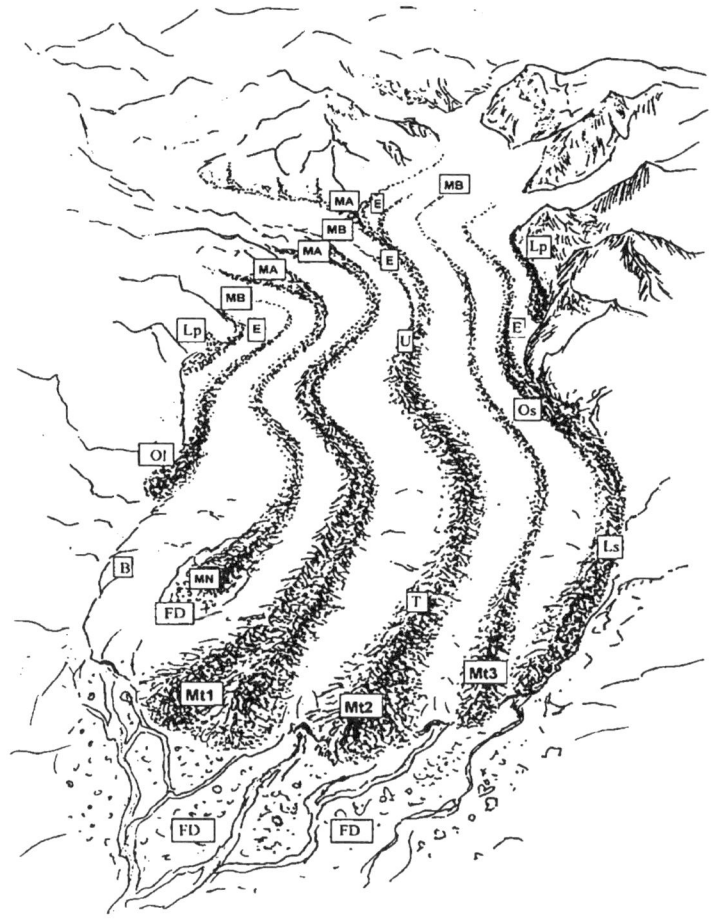

In June 2004, the Zürcher Vierteljahrsschrift published my short rejoinder to the work of Graf et al. of 2003, in which I proved mathematically that medial moraines of arbitrarily low thickness transport sufficient quantities of moraine material over the course of centuries to form considerable landforms.

From 2004 till 2014 – from the Zurich Article to the Book in Bern

In the following years, not much happened in the matter of the medial-moraine model. At least René Hantke and I published two papers together on medial moraines in the Bernese Oberland in 2005 and 2006. But in the back of my mind, I was already working on a richly illustrated book in which I would like to summarize and conclusively present the medial-moraine model. This would not be a profitable business either for me or for the publisher. But the publishing house of Haupt in Bern, the publisher of meanwhile three editions of the *Flora Helvetica*, was likely to offer support. After initial reservations on the part of the publisher, this proved to be the case: Matthias Haupt even made it a top priority and helped me to design the book both beautifully and richly. It was published in 2014 and targets laypeople, but in its final part it presents the medial-moraine model with all its consequences and, thus, diametrically contradicts the current doctrine, according to which the ground moraine is the main form of glacial deposits. It literally "reverses the lowest to the highest."

To be sure, there was another bitter disappointment for the time being: the new editor of the "Mitteilungen der Naturforschenden Gesellschaft in Bern," Dr. Thomas Burri, reviewed the book in-depth in the 2014 volume, praising the beautiful presentation and good explanations of the origin and effect of glaciers, but completely rejecting my account of the geomorphological significance of medial moraines, based on the rejection in Zurich by a mere eight geologists. This total rejection was devastating for my book and prevented a breakthrough of the medial-moraine model in Bern for another few years. As main argument against the medial-moraine model, Thomas Burri stated that the principle of actualism was not met, which requires that regular processes, which we assume for the past, also have to be verifiable in the present – and vice versa. That this was not the case with the medial-moraine model as no permanent structures formed by medial moraines could be found in today's glaciers.

SW Greenland at about 66°10'N and 49°45'W with a 1.1 km wide glacier snout. The medial moraine exhibits a wedge-like widening at its end and delivers its debris to a terminal moraine. © Google Earth, Maxar Technologies, U.S. Geological Survey

If the principle of actualism were not met, this would indeed be a devastating counter-argument. But in this respect, too, there was no danger as, in my publications, I had already pointed to various cases of permanent medial-moraine formations in today's glaciers in Valais. Now Prof. Dr. Dieter Kelletat of the University of Giessen entered the scene, an internationally experienced physiogeographer. When he, unfortunately only in 2016, got to see my book and heard editor Thomas Burri requesting readers to convince themselves of the untenability of the medial-moraine model on today's glaciers, he set to work with the help of Google Earth and, in a short time, found in glaciers on mountains all over the world an almost arbitrary number of permanent medial-moraine structures from the Little Ice Age. Thus, he proved that the principle of actualism regarding the medial-moraine model was perfectly met.

The Present State of Affairs

So where does the medial-moraine model stand today? It is a strange case in the history of science: on the one hand a tragicomical one, on the other hand an almost amusing one. Laymen, who get it explained and demonstrated in the field, understand it at once, it is so simple to understand and plausible. Even among geographers, I believe, there are quite a few who have long recognized and accepted its validity. Among the Bernese geologists there was, in the beginning, as shown above, quite an agreement, even downright relief concerning the interpretation of the Hühnerbühl. The medial-moraine view would have been accepted very quickly, had it not been for the case of the Knonauer Amt in the Canton of Zurich. The opposition that arose there was only justified to the smallest extent. But it has grown to such a extent that the medial-moraine model was pronounced dead. But it did not die, it survived the turmoil. However, it has not yet been able to establish itself among leading experts. Christian Schlüchter remains silent, since he was a co-signatory of the Zurich publication by Graf et al. Of those eight authors, none so far has distanced himself from their radical rejection.

So there is still no talk of enforcing the medial-moraine point of view. Good things take time. To generously deal with the factor of time is part of occupying oneself with earth sciences anyway. Around 1830–1860, the ice-age theory itself needed several decades until it was generally accepted. But Dieter Kelletat was probably right when he wrote to me on February 5 of this year regarding a text submitted by him to the new editor of the Bernese Mitteilungen: "I am curious what fate awaits my modest manuscript in the end, but I do not worry about the survival of medial moraines, they take care of themselves."

Publications on Medial Moraines by Gerhart Wagner and René Hantke

"Die eiszeitlichen Moränen von Aare- und Rhonegletscher im Gebiet des Worblentals bei Bern," *Mitt. Naturf. Ges. Bern* N. F. 43, 1986, pp. 63–110.

"Eiszeitliche Mittelmoränen im Berner Mittelland," *Mitt. Naturf. Ges. Bern* N. F. 54, 1997, pp. 91–137.

"Wie das Worblental eisfrei wurde. Glazialmorpholgische Beobachtungen und Deutungen nordöstlich von Bern," *Mitt. Naturf. Ges. Bern*, N. F. 56, 1999, pp. 47–76.

"Mittelmoränen eiszeitlicher Alpengletscher in der Schweiz," *Eclogae Geol. Helv.* 94, 2001, pp. 221–235.

"Mittelmoränen historischer und prähistorischer Gletscher im Wallis. Eine morphologische Studie," *Mitt. Naturf. Ges. Bern* 58, 2001, pp. 63–96.

"Drumlins im Berner Mittelland? Eine begrifflich-morphologische Studie," *Mitt. Naturf. Ges. Bern* 58, 2001, pp. 97–114.

"Eiszeitliche Mittelmoränen im Kanton Zürich, 1. Teil: Gebiet des Linthgletschers in der Zürichsee-Talung und im Knonauer-Amt," *Vjschr. Naturf. Ges. Zürich* 147/4, 2002, pp. 151–163.

"Die Eiszeitlandschaft im Gebiet Oberes Aaretal – Thunersee," *Jahrbuch vom Thuner und Brienzer See,* 2002, pp. 11–41.

"Eiszeitliche Mittelmoränen im Kanton Zürich, 2. Teil: Linth-/Rhein-Gletscher im Glatttal, Gletschergebiete von Reuss und Thur/Rhein," *Vjschr. Naturf. Ges. Zürich* 148/3, 2003, pp. 67–77.

"Eiszeitliche Mittelmoränen: Ein vergessenes Paradigma der alpinen Quartär-morphologie," *Z. Geomorph.* N. F. 47, 2003, pp. 373–392.

"Eiszeitliche Mittelmoränen im Thurgau," *Mitt. thurg. naturf. Ges.* 59, 2003, pp. 53–84.

"Das Mittelmoränen-Modell – aus wissenschaftlicher Sicht, Duplik von Wagner auf die Replik von Graf et al., in der Vierteljahrsschrift 148 (3)," *Vjschr. Naturf. Ges. Zürich* 149/2–3, 2004, pp. 83–86.

"Ältere Berner Schotter und eiszeitliche Mittelmoränen," *Mitt. Naturf. Ges. Bern* 61, 2004, pp. 10–125.

"Eiszeitliche Mittelmoränen im Bereich des Alpenrheins," *Terra plana* 4/2004, 2004, pp. 30–35.

"Eiszeitliche Mittelmoränen im Aargau," *Aarg. Naturf. Ges. Mitt.* 36, 2005, pp. 5–25.

"Eiszeitliche und nacheiszeitliche Gletscherstände im Berner Oberland, Erster Teil: Östliches Oberland bis zur Kander," *Mitt. Naturf. Ges. Bern* 62, 2005, pp. 107–134.

"Eiszeitliche und nacheiszeitliche Gletscherstände im Berner Oberland, Zweiter Teil: Täler westlich der Kander," *Mitt. Naturf. Ges. Bern* 63, 2006, pp. 133–153.

"Sandkegel auf Gletschereis," *Die Alpen,* 8/2006, pp. 26–27.

"Alpine Schuttförderbänder," *Die Alpen,* 10/2009, pp. 62–65.

"La version alpine de convoyeurs de gravier," *Les Alpes*, 10/2009.

Gerhart Wagner, René Hantke, and Dieter Kelletat, *Medial Moraines of Valley Glaciers and Surviving Ice-Marginal Landforms,* Google Earth Research 2018.

Several major articles on the subject of medial moraines have been published since 1982 in the Bernese daily newspaper *Der Bund*, first by Gerhart Wagner, later by Dölf Barben.

There are also a number of small publications by René Hantke and, above all, his book *Eiszeitalter*, Ott Verlag, 2011, with more than 100 examples of glacial medial moraines.

Fossil Necklace

Katie Paterson

A necklace comprising 170 fossils carved into spherical beads. It is a string of worlds, with each bead representing a major event in the evolution of life through the vast expanse of geological time. From the monocellular origins of life on Earth to the shifting of the continents, from the extinctions of the Cretaceous period triggered by a falling meteorite to the first blooming of flowers, *Fossil Necklace* charts the development of life on our planet.

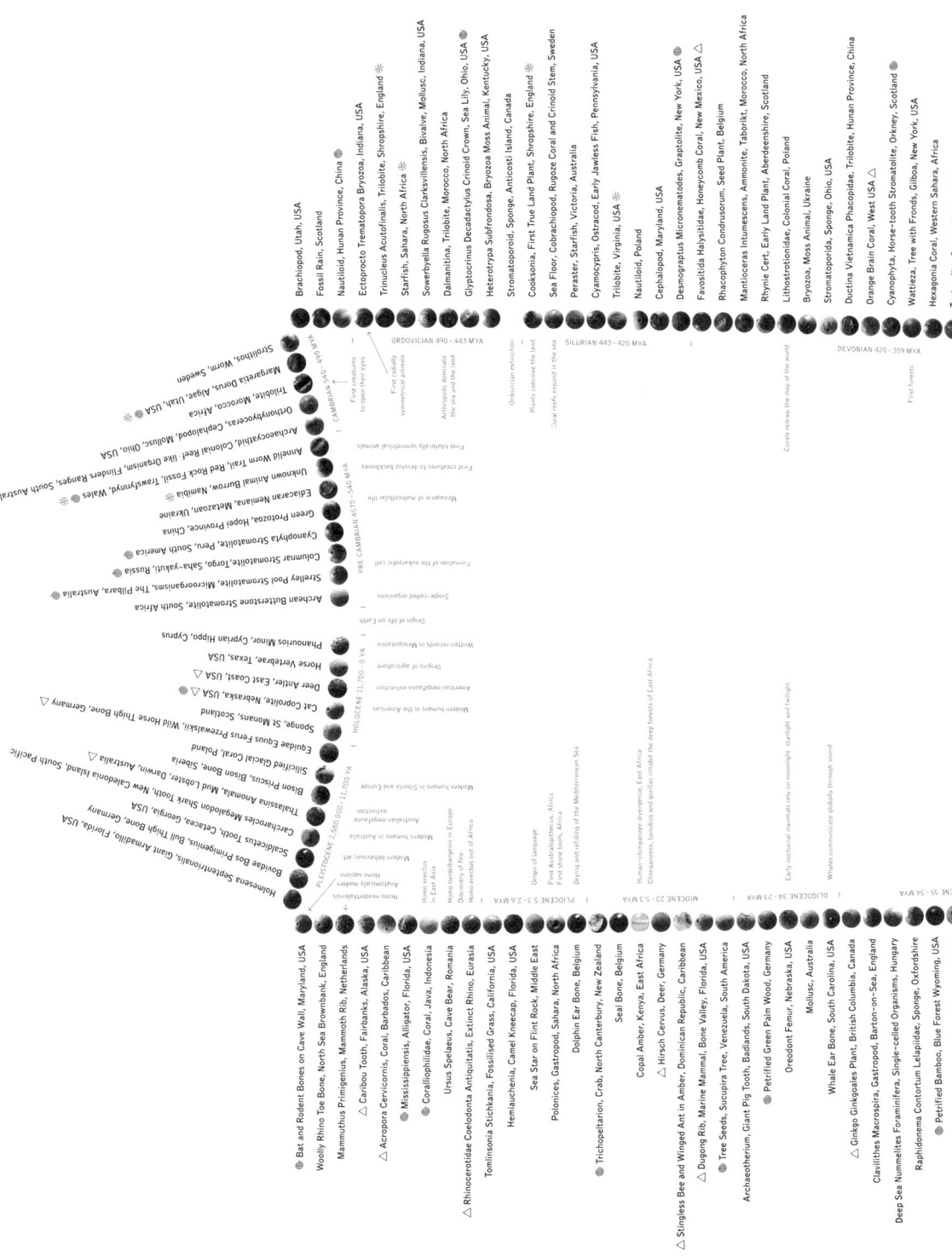

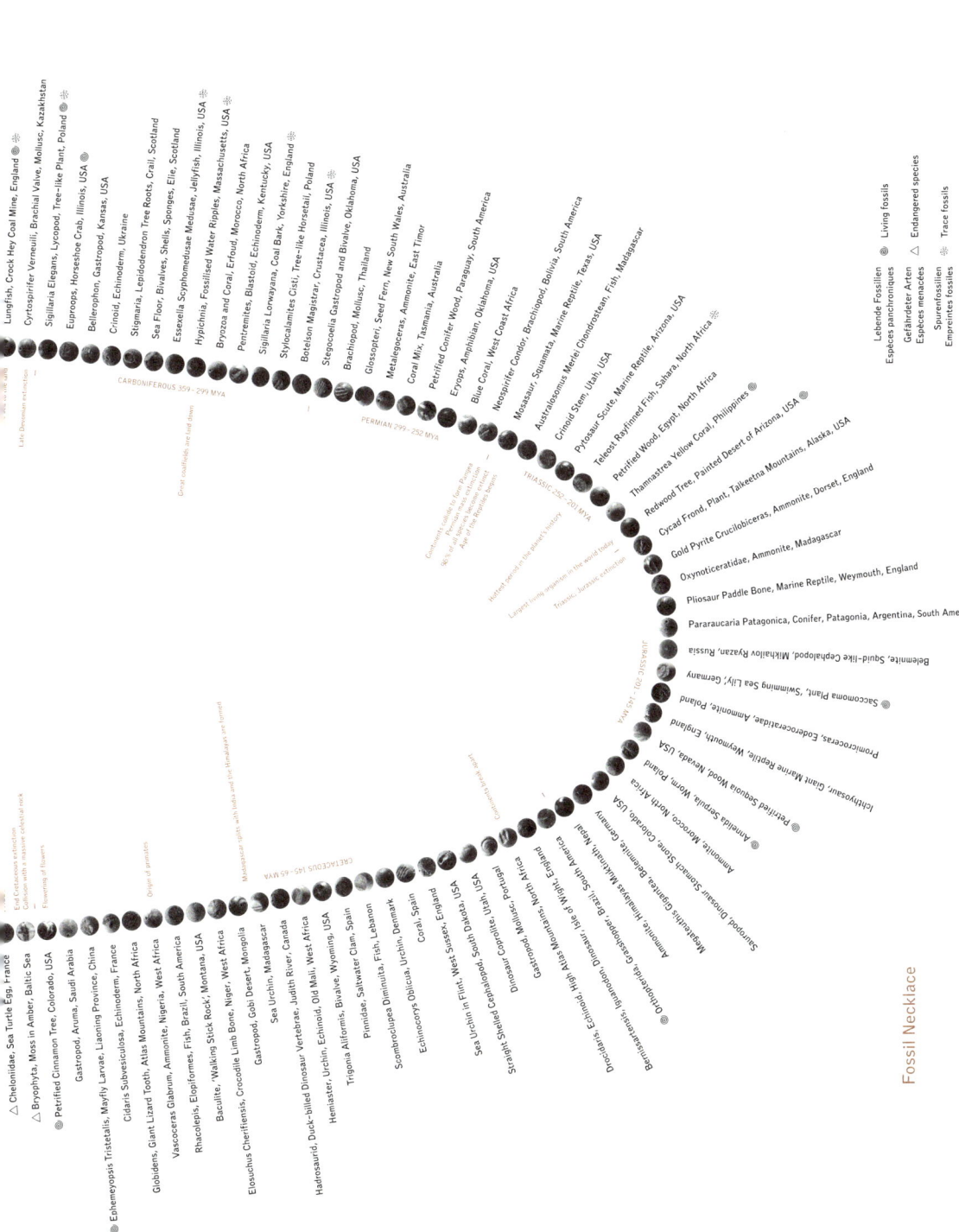

Fossil Necklace

⊛ Lebende Fossilien / Espèces panchroniques / Living fossils
△ Gefährdeter Arten / Espèces menacées / Endangered species
⁂ Spurenfossilien / Empreintes fossiles / Trace fossils

The Changing Alpine Waterscapes

Rolf Weingartner

Introduction

Commanding Mountains

Author Anais Meier recently published her first collection of stories: *About Mountains, People, and Especially Mountain Snails*. The introduction is quite provocative: "Mountains are high and mean. They are made of huge rocks, simply standing there and imposing themselves on you. Mountains are immensely egocentric and domineering. If you annoy a mountain, it will send down avalanches: in winter made of snow, in summer made of mud and scree. The little friends of the mountains are the mountain streams. They are the only ones tolerated by the mountains." And at the end of her first story, the author adds that only the sea may conceivably offend the mountains in their vanity, "because the sea clearly is much more sexually attractive than a mountain. By the sea, songs with soothing sound are written, so that all the things of the world immediately start swaying their hips."

Fig. 1: fragile Alpine region; avalanches in the Grimsel region, January 2021. Photo © Daniel Bürki

Meier paints a picture of the mountains that emphasizes their independence and uniqueness, but simultaneously points to their unpredictable nature and fragility, a fragility nota bene that will continue to increase in the course of climate change. It is this very aspect that the essay is dedicated to, in which we focus on the Alpine region – prototypically for all mountain regions of the Earth.

With the metaphor of "mountain streams are the little friends of the mountains," the narrative emphasizes the special importance of water in the mountains as a connecting element between mountains and valleys, shaping the landscape, and as a resource and source of energy. With their large variety of forms and unique soundscape, these waters significantly contribute to the harmony and beauty of mountain landscapes.

Main Features of Alpine Hydrology

The Alps play a prominent role in the hydrological cycle of Central and Southern Europe. They are a major obstacle for the atmospheric water-vapor flux, which causes an uplift of the air masses and, thus, favors the formation of clouds and precipitation. This is the reason why the mean annual precipitation in the Alpine region of 1,500 mm/a is about 1.5 to 3 times higher than in the extra-Alpine forelands. An average of 500 mm/a of this annual precipitation evaporates, which leaves 1,000 mm/a for the runoff. And this means that, on average, 1,000 l per year run off from each square meter. Truly a lot of water and the reason why the Alps are called the water tower of Europe from the point of view of hydrology and water management. The same as in a water tower of the public drinking-water supply, considerable volumes of water are located at a high altitude and, following gravity, flow to the valleys and forelands, where they are used.

Now, let us consider the seasonality of the outflow of the Rhône, a characteristic Alpine watercourse (Fig. 2): Theoretically, the runoff (R) can be determined from the difference between precipitation (P) and evaporation (ET), i.e., $R = P - ET$ (climatic water balance). The monthly runoff resulting from this equation is shown as a green line in the figure. In the winter months, the runoff is relatively

Fig. 2: mean monthly discharge of the Rhône at its inflow into Lake Geneva. Seasonal behavior of various parameters and processes. Surface of the catchment: 5,238 km², mean areal altitude: 2,127 m a.s.l. Glaciation: 11% (2020). The red line shows the mean conditions around 2085 in an emission scenario without climate protection. Illustration © Rolf Weingartner, data: Federal Office for the Environment and Hydrological Atlas of Switzerland

Runoff [mm/month]

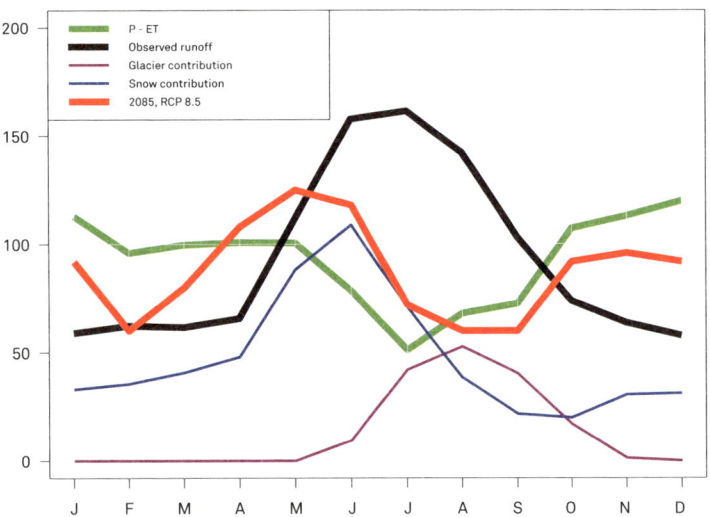

high because evaporation (ET) is very low. In summer, on the other hand, the monthly runoff is less pronounced, although the monthly precipitation is actually higher than in winter. This is due to the high evaporation rates in summer, as illustrated by the climatic balance for the month of July: A mean areal precipitation of 120 mm is balanced by an evaporation of 70 mm, resulting in a runoff of 50 mm. It is noteworthy, that this "theoretical" runoff is three times less than the one observed (Fig. 2). This difference can be explained by snow and glaciers as central and temperature-sensitive elements of the Alpine water regime: the intermediate storage of precipitation in the form of snow in the cold winter months with negative air temperatures results in very small monthly runoffs. In contrast, snowmelt between April and July and glacier melt between July and September in case of high temperatures lead to very large monthly runoffs (cf. "snow and glacier contribution" in Fig. 2). For the Rhône, about half the total discharge comes from snow melt, about 15% from glacier melt and only about a third from precipitation. These proportions are characteristic for glaciated Alpine catchments. But even in the case of the Rhine in Basel, which drains Northern Alpine Switzerland and has its sources in the High Alps, the snow content still is around 40%, while the glacier is just 1–2%.

These examples prove that snow, with its large-scale expansion, is far more important for the Alpine water balance than the glaciers, whose hydrological influence is above all quite significant in the glaciated catchments.

The non-Alpine forelands strongly depend on inflows from the Alps in summer – south of the Alps more pronouncedly than north of the Alps. Alpine meltwaters make substantial contributions to agricultural irrigation, serve to cool industrial processes, generate power in run-of-river plants and provide the base flow necessary for aquatic ecosystems.[1] The hydrological zone of influence of the Alps is, thus, much larger than the Alpine area defined on the basis of topography, such as the following example of the Rhine shows: due to its high "discharge capacity," the Alpine subcatchment contributes a mean 34% to the overall runoff in the lower reaches. That is 2.3 times more water than is to be expected based on its surface. In June, even more than half the water (52%) originates in the Alpine region.[2]

Alpine Waterscapes in an Increasingly Warm Atmosphere

Metamorphoses change the shape and/or properties of objects. The Alpine region is characterized by two major temperature-driven metamorphoses, an inner one and one on Earth's surface. 1) In the course of the folding of the Alps 135 million to 30 million years ago, hot heat flows from the Earth's interior with temperatures between a few 100°C up to around 1,000°C and, in interaction with high pressure, led to large-scale rock deformation and the transformation of the mineralogical composition of the rocks; thus granites, for example, turned into gneisses. 2) At present, we are confronted with a second major metamorphosis. This is caused by man-made greenhouse-gas emissions, which have so far led to an increase in the annual mean temperature in the Alpine region of more than 2°C since the middle of the nineteenth century. There are two main reasons for this temperature increase, which is twice as high as the global average:

1 Wilfried Haeberli and Rolf Weingartner, "In Full Transition: Key Impacts of Vanishing Mountain Ice on Water-Security at Local to Global Scales," *Water security* 11, 2020, https://doi.org/10.1016/j.wasec. 2020.100074.
2 Rolf Weingartner, Josef Fürst and Karsten Schulz, "Die Alpen als Wasserturm Europas," in: *Flüsse der Alpen. Vielfalt in Natur und Kultur*, ed. Susanne Muhar et al., Haupt Verlag, Bern, 2019.

1) the warming is less over the oceans than over land. The oceans, which cover 71% of the Earth's surface, simultaneously control and dampen the increase in global air temperature. The Swiss Alps, as a region far from the oceans, are thus in the range of the generally stronger warming. 2) In addition, warming in the Alpine region is further increased by the snow/ice albedo feedback: a self-reinforcing process. This process is triggered by climate warming, which leads to a shrinking of the area covered by ice or snow. Snow-free surfaces-in contrast to snow and ice-absorb up to 90% of the radiated energy and, thus, heat up. This warming leads to a further decrease in areas covered by snow and ice. In this context, we speak of a positive feedback.

However, we are only at the beginning of a disturbing development. Indeed, warming in the Alpine region is progressing quite relentlessly. Thus, the annual mean temperature and, correspondingly, the mean seasonal temperature will increase by a further 1.5 to 5°C by the end of the century. The extent of this increase will depend on how well we will succeed in reducing greenhouse-gas emissions.

Climate warming is hitting an Alpine region whose central elements are particularly sensitive to temperature (see above). The decrease of snow depths and the period with a snow cover, the melting of glaciers and permafrost as well as the rising zero-degree limit do not only change the hydrological processes, but also form the starting point of subsequent processes with complex interactions and, quite often, far-reaching effects on nature, society, and economy. From a hydrological point of view, three processes in particular need to be addressed:

1) The Liquefaction of the Water Balance
With the temperature increase in winter, the proportion of precipitation that is temporarily stored in the form of snow and ice decreases. The transfer of precipitation to runoff now occurs more frequently directly, i.e., without intermediate storage as snow. As a result, on the one hand, runoff increases in winter (Fig. 2) while, on the other one, less snow is available for the spring melt.

3 Hydrological Atlas of Switzerland, https://hydromapscc.ch/#en/9/46.8311/8.1931/bl_hds/NULL, Thema L04.

In addition, snowmelt begins and ends earlier in the year. Overall, the damping effect of the snow cover decreases, i.e., the snow liquefies.

The glaciers are liquefying, too, and their mass balance is becoming increasingly and gradually more strongly negative. That is, more ice melts in the course of summer than is formed in winter. This is the reason why glaciers loose volume and surface area. The degree of glaciation in the catchment of the Rhône to Lake Geneva, for instance, is currently 11%. By the end of the century, it will be reduced to about 7% in an emission scenario with climate protection and about 1% in one without climate protection.[3] The consequences of these liquefactions are multiple and conflicting: in recent decades, summer runoff from glaciated areas has sharply increased due to a severe glacier melt. Beyond a certain critical glacier volume, however, glacier melt is so low that its contribution to summer runoff is no longer significant. When exactly this critical volume is reached, depends on the size of a glacier. In many catchments with small glacier shares, this threshold is currently being reached or, by now, has already been exceeded. From this point on, summer runoff decreases sharply (Fig. 2), replacing winter as the season with the lowest monthly runoff. The summer drought becomes a new challenge. This transition from water abundance to summer deficiency will happen within one or two decades.

With climate warming, the discharges (red line in Fig. 2) will move further and further away from those observed today (black line) and increasingly approach the green

Fig. 3: The liquefaction of the water balance also affects the tourism industry. On the Diavolezza (Grisons), attempts are being made to stop the melting of the last remnant of a glacier with a fleece, thus enabling an early pre-winter start to the ski season. Photo © Rolf Weingartner

line, which is based on the climatic water balance (P - ET) and thus excludes any influence of snow and ice.

The liquefaction of the water balance does not only change the basic conditions of Alpine water management, but also worsens the water supply of the summer-dry extra-Alpine areas. But not only that. Already in 2005, Barnett et al.[4] pointed out that more than one sixth of the world's population depends on melt water from the mountains. And with climate change in mind, they warn, "the consequences of these hydrological changes for future water availability are likely to be severe."

Mass loss of glaciers in the world's mountains is also contributing to the rise of sea levels. Although the melting polar ice caps are mainly responsible for the total rise of 26.5 mm between 1961 and 2016, mountain glaciers also account for about 20%.[5] The hydrological changes in mountain areas, which cover about one-fifth of the Earth's land surface, even take on global relevance regarding global warming.

The liquefaction of the water balance in the Alpine region changes the seasonal runoff behavior and poses new challenges for water use. However, the existing artificial reservoirs and those to be built offer an interesting option for adapting to these changes. They can be used to store the increasing runoff of the winter months and make it available for the more frequently dry summers. However, this requires a reorientation of reservoir use and, thus, a paradigm shift in the political, social, and economic handling of water in the Alpine region. While, today, artificial reservoirs are typically used solely to generate power, they could be put to multiple uses in the future so that other user groups can also benefit from this water. With this "new" infrastructure, the basic supply of the population, the agriculture, and the economy can be ensured. Generating power will continue to be the dominant use of Alpine reservoirs in the future though, which is justified in view of the CO_2-free generation method, but with multiple uses they will also become part of the *public service.* The task of this public service is to ensure the basic supply of the population with a standard infrastructure.[6]

4 Tim P. Barnett, Jennifer C. Adam, and Dennis P. Lettenmaier, "Potential Impacts of a Warming Climate on Water Availability in Snow-Dominated Regions," *Nature* 438, 2005, doi. org/10.1038/nature04141.
5 Michael Zemp et al., "Global Glacier Mass Changes and Their Contributions to Sea-Level Rise from 1961 to 2016," *Nature* 568, 2019, https://www.nature. com/articles/s41586-019-1071-0.
6 *Service public* or public service is a term commonly used in Switzerland to denote a "politically defined basic supply of infrastructure goods and services, which are to be available to all population groups and regions of the country according to the same principles and in good quality and at reasonable prices" (report by the Swiss Federal Council, https://www.fedlex.admin.ch/ eli/fga/2004/767/de).

The liquefaction of the water balance affects the inhabitants, tourism industry, agriculture, ecosystems, landscape, hydropower uses, and extra-Alpine lowlands. These impacts will now be examined more in-depth using two examples.

1. Tourism: an analysis by Josi[7] of skiing in Switzerland shows that 43% of the winter-sports destinations very strongly depend on snow, while another 28% do so quite strongly and, accordingly, are affected by the ever-shorter periods with snow cover. In the Swiss Alps, the duration of the snow cover in winter has decreased by about a month between 1970 and 2015.[8] Ski resorts are trying to counteract this with all-weather snowmaking. 3,600 km, about half of the slopes in Switzerland, are already covered with technically produced snow. Investments of around CHF 3,600,000,000 have been made so far, i.e., an average of CHF 1,000,000 per kilometer of piste. In addition, there are considerable operating and maintenance costs: the operation of a large ski area generates costs in the order of CHF 250,000 per day.[9] Technical snowmaking alone accounts for 15–20% of these costs. Therefore, the economic consequences of climate change are considerable.

2. Extra-Alpine lowlands: Despite the temperature-driven changes in seasonal water supply described above and thanks to the little change in long-term mean annual runoff volumes, the Alps will not lose their function as a water tower; on the contrary, they will even gain in importance in view of the increasingly warmer forelands turning drier in summer. As a consequence, the pressure on the use of Alpine resources will increase.

2) The Destabilization of the Alpine Space

Along with climate warming, certain dangerous natural processes are changing and intensifying in the Alpine region. Often, areas that were previously considered "safe" are affected. Largely responsible for this is the rise in the zero-degree line. A 1°C increase of temperature raises the zero-degree line by 150–200 m. Man-made temperature leads to the zero-degree line being higher than today, both in all seasons and during individual flood events.

7 Pascale Josi, *Wassermanagement in der technischen Beschneiung und Zukunftsvisionen des Schneesports in der Schweiz*, MA thesis, University of Bern, 2018.
8 Geoffrey Klein et al., "Shorter Snow Cover Duration since 1970 in the Swiss Alps Due to Earlier Snowmelt More than to Later Snow Onset," *Climatic Change* 139, 2016, doi.org/10.1007/s10584-016-1806y.
9 https://www.seilbahnen.org/de/Branche/Statistiken/Fakten-Zahlen.

Fig. 4: debris flow of August 22, 2005, in the Rotlaui, Guttannen (Bernese Oberland). Photo © Beat Kehrli

This has far-reaching consequences. The increase in the summer zero-degree line affects the permafrost areas that are, depending on exposure, in the Alps 2,500–2,700 m a.s.l. Permafrost occurs when temperatures in the sub-surface below the low-density active layer, which thaws in summer and freezes again in winter, do not rise above 0°C. The permanently frozen bedrock provides stability to the rock, scree slopes, and moraines. As the climate warms, the summer zero-degree line reaches the area of permafrost, which begins to thaw. With the thawing of the permafrost, this stability is gradually lost. Landslides, debris flows, and rockfalls are the result. Debris flows are particularly devastating because they usually reach the valleys and endanger villages and infrastructures, as shown by the example of the largest debris flow in the Rotlaui in Guttannen (Bernese Oberland) in recent decades (Fig. 4). This debris flow transported more than 500,000 m³ of material[10] to the valley floor. There, not only the Grimsel Pass road was flooded, but the valley river, the Aare, was first dammed back and then, after over-flowing the damming debris-flow embankments, led over the Grimsel Pass road directly into the village of Guttannen, where large floods caused major damages.

Also, for flood events in the Alpine region, the position of the zero-degree line significantly influences the process: with climate warming, the probability increases that the zero-degree line for heavy precipitation events will be higher than today. With a higher zero-degree line, however, the proportion of a catchment that is over-rained increases because the precipitation falls in liquid form up to high elevations and, therefore, directly contributes

10 To illustrate this volume: Forty soccer fields could be covered by two meters of rubble each.

to flood runoff, i.e. cannot be temporarily stored in the form of snow. At the same time, the thawing of the permafrost and the melting of the glaciers will expose debris and scree that will be mobilized during rainfall up to high altitudes, causing an increased bed-load discharge in streams and rivers. If you also consider that the intensity of heavy precipitation will increase by 7%[11] with a temperature increase of 1°C, one thing is obvious: the flood risk in the Alpine region will increase and require adjustments of technical and spatial planning.

3) The Graying of the Landscape

Natural and cultural diversity is characteristic of mountain regions. It develops due to the Alpine altitudinal gradients and the associated changing vegetation and land use, due to the small-chambered relief, due to differences in exposition and slope inclination, and due to contrasting climatic, pedological, and geological conditions. As Bätzing[12] so aptly writes, most people see and experience the Alps with their glaciers, rocks, forests, meadows, and pastures as a magnificent, vast, and intact natural landscape. Therefore, the Alps were and continue to be identity-forming, as when today's Swiss Confederation was established in 1848. In the present day, the Alps form a natural counterworld to the surrounding forelands, which is increasingly suffering from settlement pressure and densification. The Alpine image of Switzerland continues to be cherished and conveyed to the world. Guests arriving at Zurich Airport, for example, are already introduced to Switzerland as an Alpine country on the commuter train to the main terminal by tinkling cowbells and popular "Alpine" music.

Settlements, hydropower plants, reservoirs, mountain roads, and high-voltage power lines, etc. are not conducive to this image of the Alps, but cannot destroy the image of an "ideal" world, an Alpine Arcadia. Also, bear in mind that the more you focus on the High Alps, the closer you come to the idea of an intact nature.

Along with climate change, the resulting disappearance of the glaciers and the decrease in snow cover, the landscape of the High Alps also changes. It is characterized by the grandiose and incomparable interplay between the

11 Clausius-Clapeyron's law.
12 Werner Bätzing, *Zwischen Wildnis und Freizeitpark – Eine Streitschrift zur Zukunft der Alpen*, Rotpunktverlag, Zurich, 2017.

Fig. 5: graying of the landscape, Diavolezza, Val Bernina, August 2018. Photo © Rolf Weingartner

white of snow and ice and the gray hues of the rock and rubble surfaces and has inspired generations of painters since Caspar Wolf (1735–1783), the "pioneer of Alpine painting."[13] With the increase in temperature, however, the white is now gradually disappearing. Bernese geographer Paul Messerli puts it in a nutshell: "The sublimity disappears and what remains are gray hues"[14] (Fig. 5). This has an image emerge in which the threat menacing it comes once again to the fore. "When I walk through the Alps today, I am interested in other phenomena than thirty years ago. The development of ski lifts is less of a problem than the retreat of the glaciers," thus Messerli.

In the grayed landscapes, due to the melting back of the glaciers, glacially eroded topographic cavities are exposed, which rapidly fill with water. The total volume of these approximately 500 to 600 cavities is about 2 km^3, which corresponds to about 3% of the present glacier volume of Switzerland.[15] For comparison, the usable volume of all Swiss reservoirs is about 3.5 km^3.[16] These cavities, often referred to as "new lakes," have a great ecological, scenic, and touristic potential, but are also suitable for hydropower exploitation (Fig. 6). Conflicts of interest and use seem to be inevitable!

What Is to Be Done?

Liquefied, destabilized, and graying: the Alps, and with them many mountain areas, clearly are hotspots of climate warming in the truest sense of the word. The changes are clearly recognizable to everyone, and both visible and

13 Susan Misicka, "Schweizer Alpenmaler erhält eigenes Museum," *Swissinfo*, April 6, 2019, https://www.swissinfo.ch/ger/caspar-wolf_schweizer-alpenmaler-erhaelt-eigenes-museum/44877038.
14 Bernhard Ott, "Die Menschen schaffen sich Natur in ihrer Nähe," *Der Bund*, August 8, 2017.
15 Wilfried Haeberli et al., "Gletscherschwund und neue Seen in den Schweizer Alpen," *Wasser Energie Luft* 104 (2), 2012, https://doi.org/10.5167/uzh-140414.
16 Hydrological Atlas of Switzerland, print version, Table 5.3., https://hydrologicalatlas.ch/products/printed-issue/rivers-and-lakes/plate-5-3.
17 OECD, *The Economic Consequences of Climate Change*, OECD Publishing, Paris, 2015, https://doi.org/10.1787/9789264235410-en.
18 https://www.bafu.admin.ch/bafu/de/home/themen/klima/dossiers/klimaschutz-und-co2-gesetz.html.

measurable. They confirm the forecasts of science, which has a very good understanding of current processes and those that will decide the future. In order to minimize the effects of climate change in the Alpine region, immediate and consistent action is needed, embedded in a far-sighted and coherent planning of measures. Below, we will discuss three basic, complementary ways to achieve this; a discussion that culminates in the ultimate call to finally take active action to protect the Alps.

The Silver Bullet

Avoiding greenhouse-gas emissions is the best way to stop the processes of change in the medium to long term. There is no way around this to fight the causes, absolutely none. This is the only way to reduce and then stop a further increase in temperature. Climate protection is the best protection of the Alps (Fig. 7).

Global warming will cause very high costs in the medium to long term. These costs would clearly exceed the costs of climate protection measures. The OECD puts the cost of unchecked global warming at up to 10% of its gross domestic product (GDP) by the end of the century.[17] By contrast, the measures needed to limit global warming to 2°C will be around 1% percent of annual GDP. These costs will be significantly smaller, the sooner global warming can be mitigated.[18]

Fig. 6: the formation and development of a kettle hole at the Trift Glacier in the Bernese Oberland. In 1948, this cavity was covered by the glacier. By 2006, the glacier had already melted to a considerable degree. The kettle hole underneath the glacier became visible and is now filled with water. It is now being planned to use this kettle hole as a reservoir. Photo © Sammlung Gesellschaft für ökologische Forschung, Munich (pictures of 1948 and 2006), Kraftwerke Oberhasli (image montage 2028)

1948

2006

2028

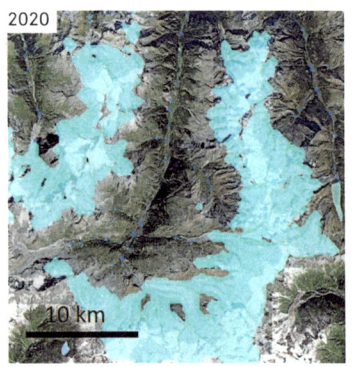

2020

10 km

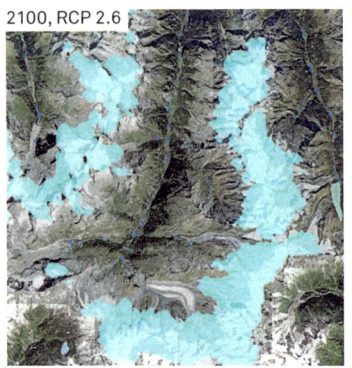

2100, RCP 2.6

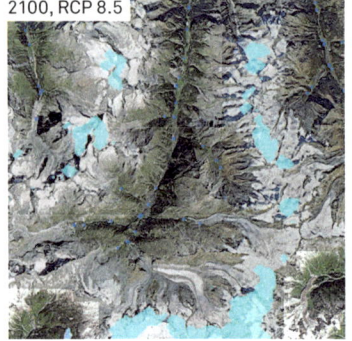

2100, RCP 8.5

Fig. 7: change in glacier surfaces (light blue) in the region of Zermatt (Valais) between 2020 and 2100 depending on the emission scenario. Hydro-CH2018: Scenarios Until 2100. Map © Hydrological Atlas of Switzerland

Against this background, it is quite incomprehensible that a majority of the Swiss population, a full 51.6 %, that is, rejected a new law in June 2021 that aimed at further reducing greenhouse-gas emissions. Apparently, it was not understood that the atmosphere is a reservoir on which our life (clean air) and the survival of future generations (concentration of greenhouse gases) depend. The exhaust and greenhouse gases fed into this reservoir – especially those that degrade only slowly – accumulate and change the properties of this reservoir with sometimes fatal effects for humans and the ecosystem. The CO_2 (carbon dioxide) emitted by humans only very slowly degrades in the atmosphere. After a thousand years, between 15% and 40% are still left.[19]

In order to reach the 2°C climate target set in Paris, a maximum of 420 to 1,070 gigatons of additional CO_2 may be emitted. Currently, emissions amount to 42 gigatons of CO_2 per year and are rising. In a worst-case scenario, less than ten years remain until this threshold is reached. If the latter is exceeded, the 2°C target will be lost for good, though further efforts at reduction will still be needed, otherwise the next targets (2.5°C, 3.0°C, etc.) will also be missed.

Measures of Adaptation
It is imperative that the measures to combat the causes, which range from the global to the national level, be supplemented with tailored adaptation measures at the (inter)regional and local level. For even if the international community succeeds in reducing greenhouse-gas emissions quickly and sustainably, the warming cannot be stopped immediately because the "braking distance" will take several decades or even centuries.

In Switzerland, the National Research Program NRP61 on "Sustainable Water Use in Switzerland" was conducted between 2012 and 2015, given the climate change. From its rich palette of results and findings, two points are worth emphasizing from the point of view of adaptation measures:

1. Adaption measures may not be one-sidedly and exclusively focused on the impacts of climate change. In

fact, depending on the region, the changes triggered by socio-economic change could be even more decisive for the future water situation than climate change.

2. Water problems are mainly caused by management problems, as shown, e.g., by the analyses in the Crans-Montana-Sierre region: "Since available water resources are generally large enough to cover all demands, the currently observed problems of […] water scarcity has to be attributed to regional problems of water management. Large differences were observed between municipalities: water-rich or water-poor municipalities, water demand in one municipality twice as high as in another, and very different water prices overall. This indicates a significant lack of equity at the regional level, among municipalities and among citizens. Plus, there is a lack of transparency in the costs of water management, which further reduces the sustainability of the current system. Unequal access to water is further fostered by an insistence on old water rights and a degree of legal ambiguity that impedes the development of a visionary cooperation and projects. The technical solutions developed in each separate case in the short term prevent any real political ambition to solve water management problems."[20]

These two points illustrate the need for a paradigm shift from simply using available water to managing water use. A holistic planning of measures that anticipates both climatic and socio-economic developments is needed. Such pro-active action has never been a strength of politics, in contrast to the well-functioning re-active action after major natural events, which is usually very successful in the Alpine region to find and implement quick and efficient solutions.

Avoiding Conflicts of Use

Climate change is altering the Alpine region, increasing the pressure on its resources, but also creates interesting options with the "new lakes," which have great ecological, scenic, and tourist potential (Fig. 6). As a consequence, the conflicts of use that already exist today will continue

19 Mercator Research Institute on Global Commons and Climate Change, "Klimawandel: Wieviel CO_2 bleibt uns noch?," *Scinexx*, November 12, 2018, https://www.scinexx.de/news/geowissen/klimawandel-wie-viel-co2-bleibt-uns-noch.

20 Rolf Weingartner et al., *MontanAqua: Wasserbewirtschaftung in Zeiten von Knappheit und globalem Wandel – Wasserbewirtschaftungsoptionen für die Region Crans-Montana-Sierre im Wallis*, Forschungsbericht des Nationalen Forschungsprogramms NFP 61, Bern, 2014, http://www.nfp61.ch/SiteCollectionDocuments/nfp61_projekt_montan-aqua_broschuere_d.pdf.

to intensify, especially regarding the water supply. Accordingly, it is important to initiate an Alps-wide and cross-sectoral clarification process regarding the future handling of water as a resource and, thus, develop and introduce a clear formal framework, which so far exists at most in rudimentary form. "This is not about perpetuating existing uses and use claims, but about establishing a new coexistence of these uses; for instance regarding the so-called priority areas, which allot certain uses a high-priority status to allow them to develop their potential ahead of anything else."[21] Figure 8 provides a (fictitious) example of this.

Action at Last

On May 11, 1990, 32 years ago, an international symposium was held at ETH Zurich on the topic of "Snow, Ice, and Water of the Alps in a Warmer Atmosphere."[22] At that time, climate change was still a vague concept, as the following quotation shows: "[The observed temperature increase of 0.5°C] could already include an anthropogenically enhanced greenhouse effect." Nevertheless, the decisive processes of change in the Alpine region were already correctly understood at that time: "According to the best currently possible estimates, a temperature increase of 3°C would shorten the duration of snow cover at resort altitude by about one month and, at the same time, reduce snow depth by about half; three quarters of today's glacier areas would disappear and many permafrost slopes below 3,000 meters would be destabilized in the long term."

And in the panel discussion at the meeting, Kerry Kelts, then director of Proclim,[23] made the following comments: "The causal importance of climate, as well as the irreversibility of climate change, poses a serious problem. Globally, it would mean a great deal to reverse the trend. It would be necessary to have a go at it with all our strength, without waiting for further evidence. This political activity will undoubtedly not happen very quickly. What matters is understanding the processes. The next 15 years will be crucial as to the measures to be taken."

21 Thomas Kissling et al., *Profilierung des Alpenraums, Projektskizze,* ETH Zurich internal report, 2021.
22 *Mitteilungen der VAW* 108, https://ethz.ch/content/dam/ethz/special-interest/baug/vaw/vaw-dam/documents/das-institut/mitteilungen/1990-1999/108.pdf. The following quotes are taken from this VAW notification.
23 https://proclim.scnat.ch/de.

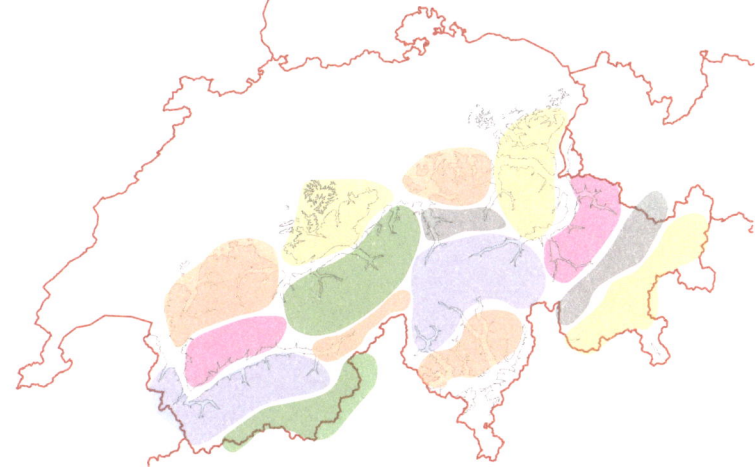

Fig. 8: fictitious map sketch for a super-ordinate planning of the alpine space in Switzerland. 2021, Map (source: geo.admin.ch) © Chair of Günther Vogt, ETH Zurich

Projection

🟨 Cooperation areas
🔴 Priority areas tourism
🟪 Priority areas hydropower
🟢 Priority areas nature conservation
🟠 Priority areas agriculture
⚫ Vernacular areas

However, in the meantime more than thirty years have passed. For the Alpine region, too, science has now produced a comprehensive quantitative understanding of the effects of climate change, differentiated in space and time. Moreover, their statements have remained consistent over the entire period. This is also due to the fact that in the Alpine region – as has already been mentioned several times – the decisive processes are controlled by the prevalent temperatures. That's why the direction of the changes has been understood very well for quite a long time now and with more and more detailed models also their timing and intensity depending on the emission scenarios. In spite of this, we do not see adequate action! Actually, we have been observing as contemporary witnesses for a mere three decades now, meticulously recording the changes already taking place and reporting them in detail. The crucial question though is whether the step from "observing" to "acting" will be taken?

A major reason for people's lack of willingness to act is that climate change is actually a classic problem of the commons.[24] The introduction of greenhouse gases changes the basic functions of the atmosphere in such a way that the life of us all is endangered. Hardin, who coined the term "tragedy of the commons"[25] to define the problem, described in 1968 how individual actions to assert one's own interests have devastating consequences for the collective. Basically, every individual knows that uncooperative behavior harms the entire community.

24 Eleonore Ostrom, *Governing the Commons. The Evolution of Institutions for Collective Action*, Cambridge, University Press, 1990.
25 Garrett Hardin, "The Tragedy of the Commons," *Science* 162, 1968, https://science.sciencemag.org/content/162/3859/1243.

Nonetheless, no one wants to restrict himself, only to find out how the others profit by further exploiting the resources in a selfish way, i.e., in the case of the atmosphere, by further polluting it with greenhouse gases. "Freedom on the commons," Hardin concludes, "ruins all those involved."[26]

The German Advisory Council on Global Change (WBGU)[27] outlines an interesting way out of this dilemma with the model of the "shaping state." In this context, the state becomes the decisive "actor of transformation" with competencies that go beyond the traditional regulatory function in order to help the necessary turnaround in climate protection achieve a breakthrough. The WBGU speaks of a "social contract for a Great Transformation." For such a contract to come into being, it is however not enough to continue with existing policies; what is needed is a paradigm shift or, as the younger generation puts it, a system change. The shaping state sets clear guidelines with laws, incentives, incentive taxes, prohibitions, and other suitable instruments, provides the necessary funds and creates a binding framework for the necessary concrete action at regional to local level.

Boundaries Are Shifting

With climate change, boundaries in the mountain areas are changing: zero-degree line, snow line, equilibrium line for glaciers, lower limit of permafrost, beginning and end of ski seasons, and the boundaries of hazard areas. With the liquefaction of the water balance, not only the seasonal runoff patterns change, but Alpine landscapes, too (Fig. 9). The latter are graying, opening up new insights and prospects. The Alpine image is changing from within.

In the nineteenth century, the Alps became less important as peripheral regions on the outer borders of the newly emerging nation states and far away from the political and economic centers. However, in view of the climatic and socio-economic challenges in the extra-Alpine areas, they will now distinguish themselves as resource, energy, and recreation region, in short; as an ecological island in Central and Southern Europe. At the same time, pressures and desires from outside the region will increase. Therefore, a new coexistence between Alpine and extra-Alpine

26 https://www.wikiwand.com/de/Tragik_der_Allmende
27 WBGU, *Welt im Wandel – Gesellschaftsvertrag für eine Grosse Transformation,* WBGU, 2011, https://www.bundestag.de/resource/blob/434158/6fbf11d713565fa35d4387383389407d/adrs-18-228-data.pdf
28 Paul Messerli, "Neue Solidarität zwischen Berg und Tal," *Der Bund,* January 30, 2021.

Fig. 9: Gamchi Glacier, Bernese Oberland, on July 16, 2011 (top) and June 24, 2018 (bottom). Photo © Simon Oberli, www.SwissGlaciers.org

regions will have to be established or, in the words of Bernese geographer Paul Messerli, a "new solidarity between mountain and valley" needs to emerge.[28]

From a hydrological point of view, water in the Alpine region clearly only becomes a problem if nothing is done. The solutions are outlined, but mostly only on paper. Society, politics, and the economy will finally have to leave their previous mental framework and focus on new, innovative, sustainable and, above all, holistic solutions. Such a change does not happen within a mere few years. So, it is better to start now than put it off for tomorrow.

Towards No Earthly Pole

Julian Charrière

Glacial landscapes have never before had such a strong visual presence in popular culture, where they serve as prominent symbols of anthropogenic climate disturbance. Although few people have visited their remote geographies, glacial regions loom large in the collective imagination, as the last stronghold and melting ideal of a fantasized reality. *Towards No Earthly Pole* was conceived in 2017, when Julian Charrière was invited aboard a Russian research ship as part of the first Antarctic Biennial, inspiring subsequent research and expeditions to equally remote and inhospitable glacial regions. The project was realized not only in the polar nights but also in Alpine locations such as the Rhône, Morteratsch, and Aletsch glaciers in Switzerland, or on the slopes of Mont Blanc in France, rendering an uncanny and eerily blackened vision of the cryospheres in contrast to the vistas of dazzling landscapes of white snow and bright daylight generally depicted. A spotlight carried by a drone reveals the massive landscape in snippets, limiting the range of vision and heightening its drama by the deep shadows the icebergs cast. *Towards No Earthly Pole* demonstrates a shift in the mode of representation by delineating a place that has been overly exposed in another light. Through the use of moving artificial lighting in addition to the physical characteristics of the subject matter, it is almost impossible to gain some orientation and sense the scale of the depicted object, which results in a complete sensory loss of scale, time, and any marker allowing us to discern the natural from the virtual. Interrogating glacial mythologies and iconographies from the age of exploration to the contemporary era of the environmental crisis, the project seeks an un-grounding of Romantic ideas of nature – offering instead a techno-logically mediated visual narrative of the exquisitely sensitive ecology of high-altitude and high-latitude regions.

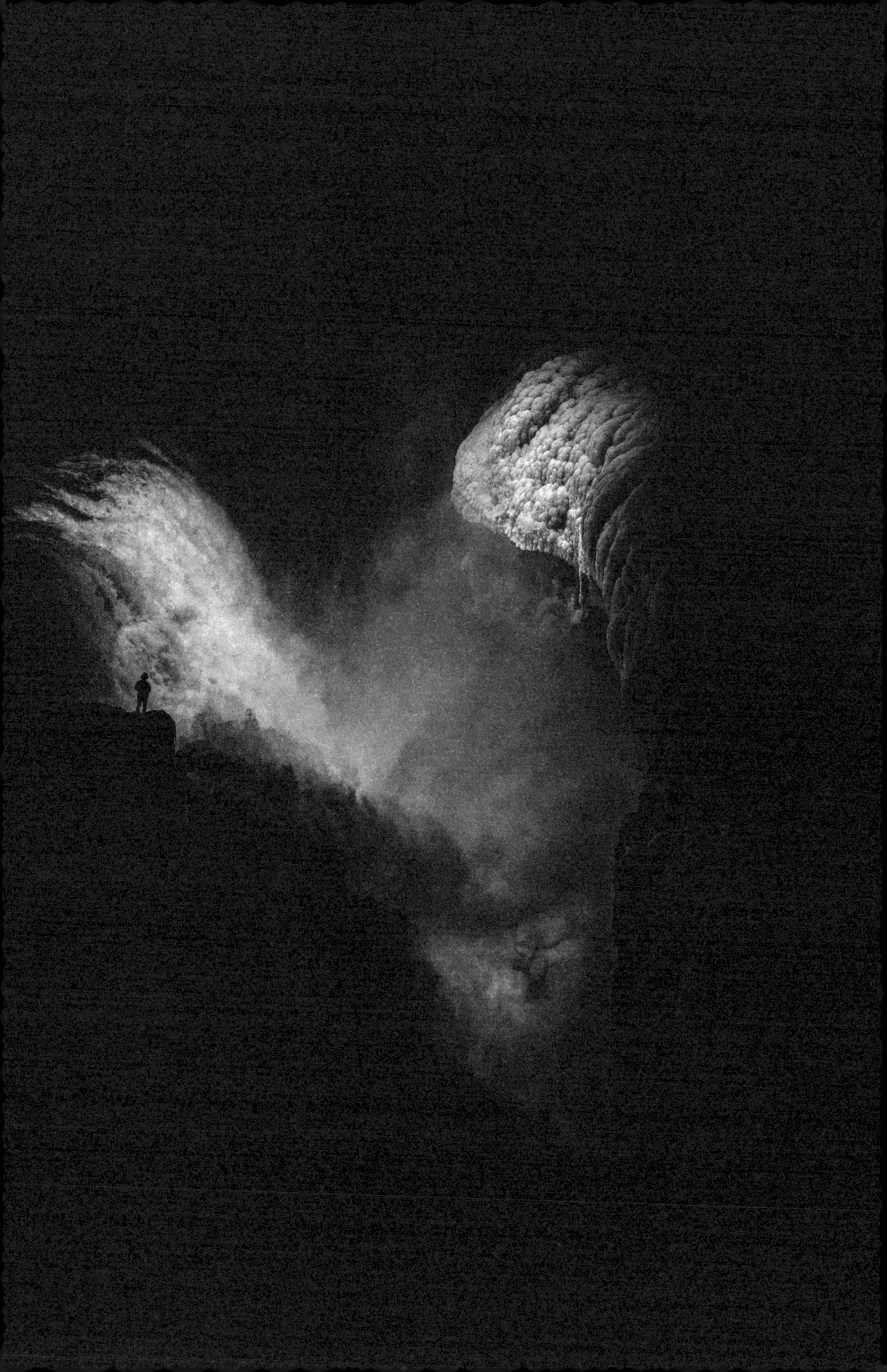

Where Was the Glacier?

Lukas Ryffel

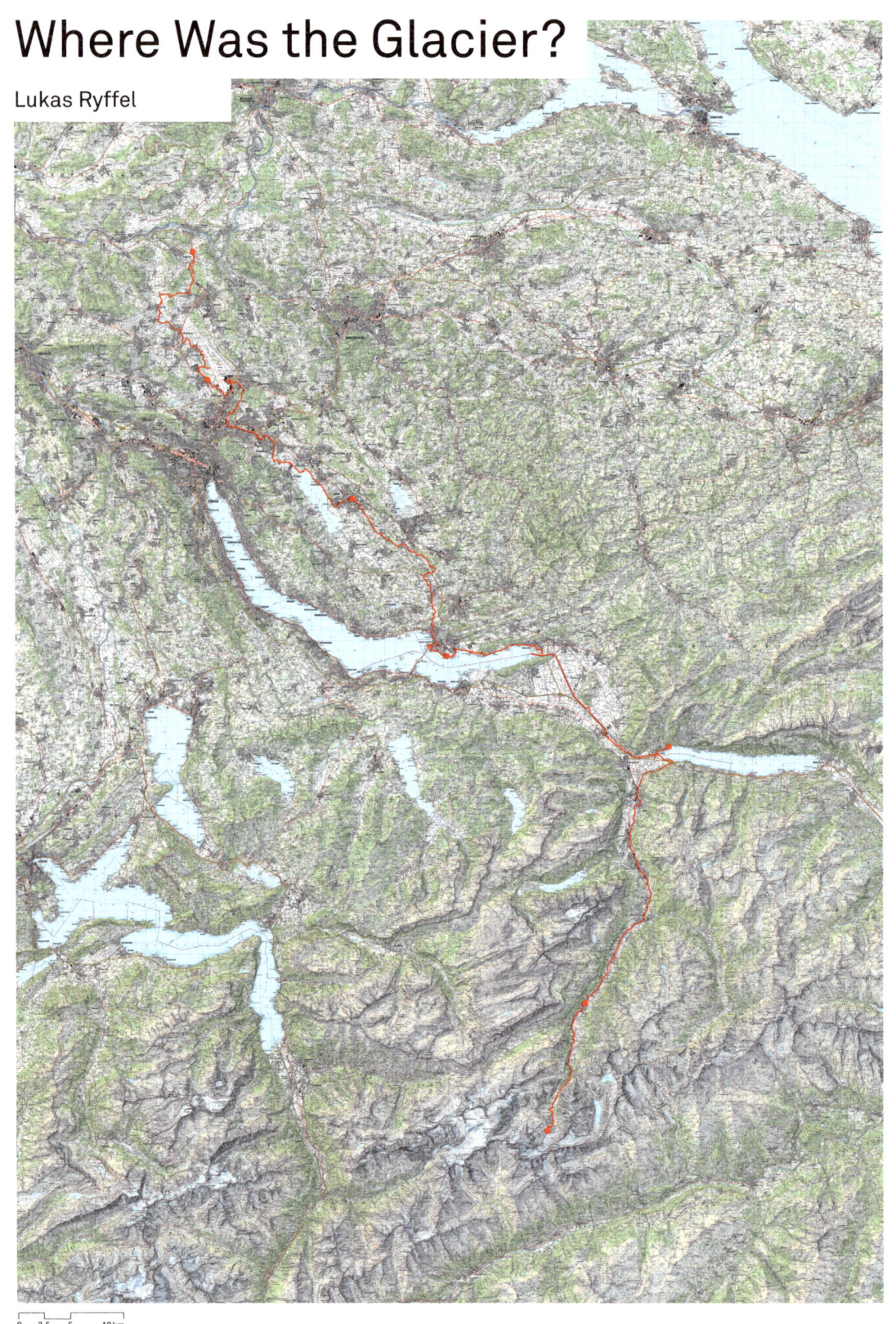

0   2.5   5   10 km

The six-day hike starting in Glattfelden takes the traveling architect far into the Alpine landscape. With an attentive eye but without specific prior knowledge, he sets off in search of glacier traces, trying to find out just how the glacier shaped the landscape and, in the process, what was built. The story is told along the traveled route in pictures and texts. In that way, space and time can be physically experienced.

Most of the places where, now, there are Swiss cities, were once covered by ice. Glaciers shaped the landscape as we know it today. They carved steep valleys out of the earth, left behind rolling hills and many lakes. I grew up in this landscape, on the outskirts of Zurich – my whole life I was surrounded by it. But never have I met the glacier, this huge thing, that is supposedly still alive. It is time to look for it.

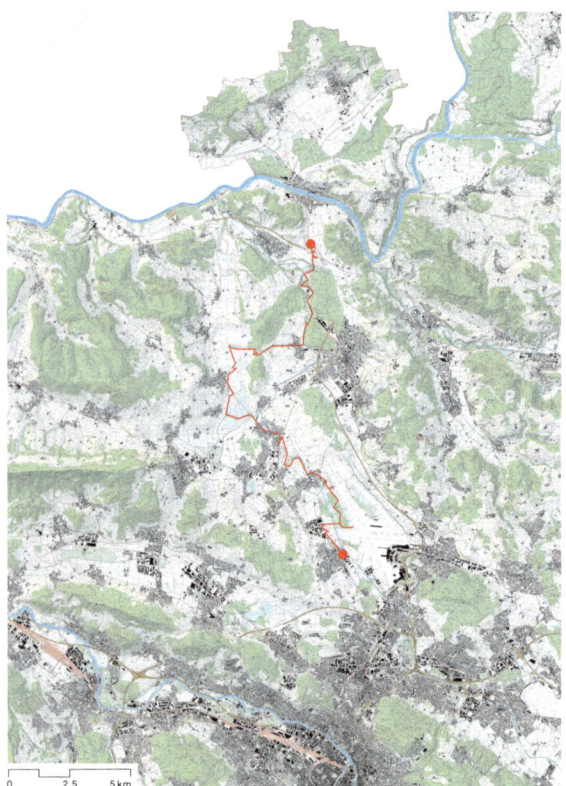

0 2.5 5 km

Day 1: Monday, June 7, 2021, 32.2 km, 8h 40min

My journey begins on a field in Glattfelden, close to the border to Germany, close to the Rhine River. According to the map "Switzerland During the Last Glacial Maximum," the glacier ended on this field around 25,000 years ago.

Cars every second. In between many trucks, all loaded with gravel, earth, or building rubble. The glacial landscape is still moving, driven around. The stones in the gravel pit behind me are round, shaped by the river. Poppies flower in red in between. Holes are dug, then filled up again. The noises of the heavy trucks make me tired, even though I just started my journey.

Butterflies dance on the riverbank. I follow the river, I follow the glacier, I walk below the ice.

Up the rolling hills that were formed during the advances and retreats of the glacier. Like on a conveyer belt, debris was carried here and accumulated in the end moraine, marking the maximum advance of the ice. 18,000 years ago, after the final thaw, a large lake was covering everything, retained by the hills. The deposition of organic material and sediments caused it to silt up, and small lakes and many peat bogs are the only traces that I can find of it.

Where the peat bogs were drained, the land is fertile. Sunflowers on one side and rapeseed on the other. In between, a continuous flow of cars. Above me, power poles, electricity crackling softly in the air. The ice cover must have been higher than the power poles.

It starts to rain heavily. I flee into a tiny tunnel that runs under a road. Built for amphibians, so they do not fall victim to the many cars. The tunnel is narrow, I sit huddled inside, outside there's thunder. There are many of these amphibian tunnels here. The peat bogs are

a habitat for various rare animals and plants, but the big roads cut their territory into pieces.

Planes fly over my head every minute. The power poles crackle even more after the rain. Nearby, cows eat grass and look at me for a long time.

Right next to the airport a nature reserve, also a peat bog. Dragonflies next to the planes. Mosquitoes bite me and a private

jet thunders past. After the Second World War, it was decided that this unproductive bog land of no agricultural use should become an intercontinental airport. The already relatively flat area was levelled, rivers were canalized and the peat bogs were drained. Today, the infrastructure of the airport and the surrounding bogs are closely connected, nature is increasingly becoming part of the infrastructure. Borders dissolve.

The day ends as it began. With trucks filled with gravel and building rubble. Right next to the airport

is a huge recycling plant that converts construction waste into gravel and recycled concrete. Here, the glacial landscape is no longer the resource, but the city.

A rounded stone, found in the gravel pit on the first day of my journey. Shaped over time, moved through the land.

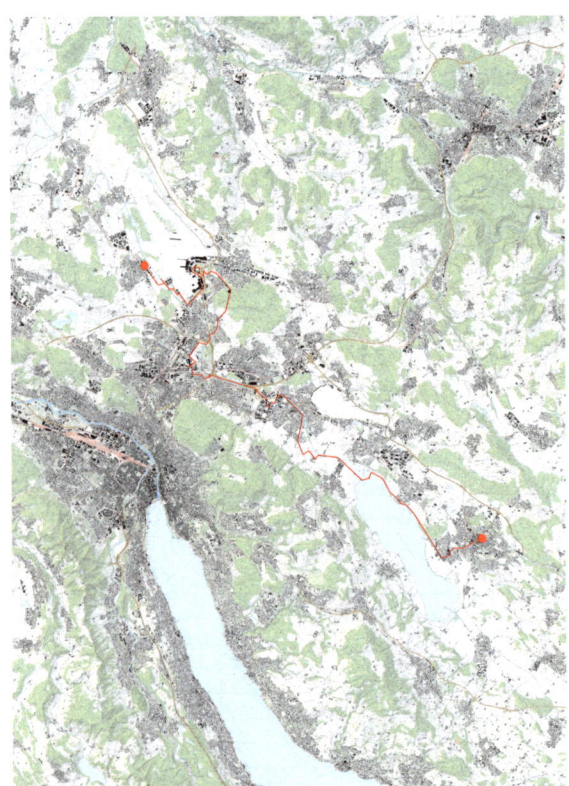

Day 2: Tuesday, June 8, 2021, 40.3 km, 9h 43min

Shopping centers, transit hotels, petrol stations, car dealers, and snack bars. On one side lies the suburban railway and on the other the airport. I walk along the grewen fence which gates it. A rollerblade and bicycle path runs along the fence. Later, many empty office buildings. Available for rent. In between industry and brothels. Concrete, asphalt, many cars, the highway, the tram on a flyover.

After a while, I stand in front of the main entrance of Zurich Airport. Everything is round, everything is curved. The streets, the flyovers, and the buildings. I took a detour to come here because the round shapes intrigued me on the map. I started to wonder if not only the location of the airport is related to the glacial landscape but perhaps all the buildings, too. And indeed, they are all snuggling tightly around a moraine hill, imitating its shape.

A highway cuts through the hill like a river. Everything is an amalgam of natural and human influences, impossible to untangle. I manage to find an underpass to get to the other side and up the larger hill behind. A radar station in the middle of allotment gardens rotates continuously.

In the forest, the birds are tweeting, from far away the sounds of the city, almost forgotten. Fields and farms, an unexpected rural feeling, right next to the city, which lies like a lake in the valley.

I reach the river again that I had followed the day before. A green canal, leading through the dense city. Next to it a highway. Parallel. Both have a bridge, both are able to cut through hills. They follow each other for a while. I follow them as well. The highway roars louder than the water.

A heron moves elegantly along the bank that separates the river from the highway. The power poles show me the way to the glacier.

Strawberries between industrial halls and residential towers. I pass by the Swiss Federal Institute of Aquatic Science and Technology. I have heard that some years ago a scientist researching here figured out that the groundwater underneath originates from before the last ice age. 27,000 years ago, water accumulated 100 m below and

ever since was not touched. But they dug holes to determine the age of the water and test its quality. I have a sip from their tap before I continue, in the hope I can taste this fossil liquid.

There's thunder. Roads and cornfields. The city of Zurich is now behind me. The moraine hills on both sides are getting flatter, the valley broader. Agriculture again. Sheep, lettuce, leeks, and rolling hills. Round stones in the fields.

A small, idyllic lake in a hollow, formed by the glacier. Two fishermen are talking together. Neither of them has caught anything yet. One of them says that when the water gets warmer, the small fish will come and so will the big ones. The other replies that he already caught a pike a fortnight ago, 45 cm big.

Far away, in the haze of the clouds, I see the mountains, the Glarnerland, my destination. My heels hurt, but there are no blisters yet.

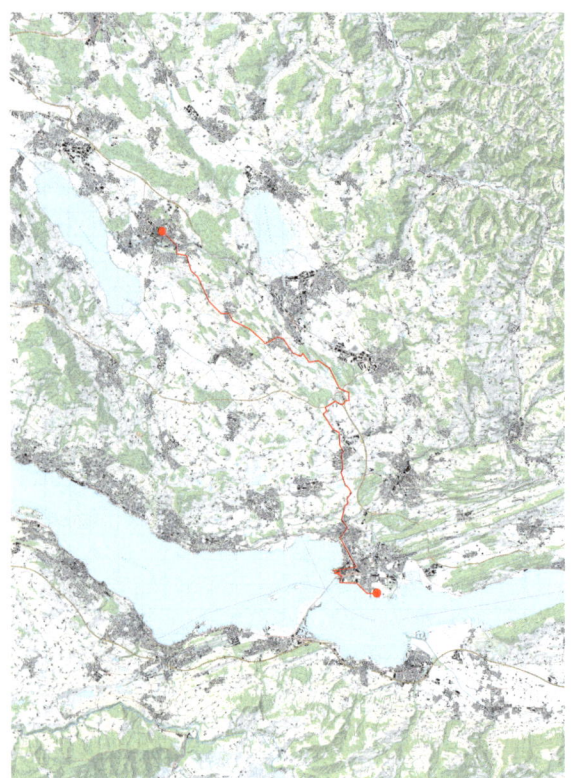

Day 3: Wednesday, June 9, 2021, 30.8 km, 10h 19min

I continue my journey from Uster, my hometown. At the end of the village, I climb a small hill. I remember how we used to fly hot-air balloons here as children on New Year's Eve. I can see over the whole valley: houses, towers and far away a glimpse of the airport. All this was filled with ice, and for the first time I get an inkling of how unimaginably large the glacier was.

The hill I am standing on is called a *drumlin*, a Celtic term for "elongated hill." Formed under the glacier, which scraped away the softer layers of the ground and sculpted the debris into drop-like shapes. They all face northwest, in the direction of the glacier flow. Compass needles in the landscape show me the direction.

The hills are gently curving, like the buildings and streets next to the airport. On top they

are mostly forested, small green hats, scattered all around. It is not possible to see into the distance anymore, always there is a hill in the way. Senior citizens on bicycles are on the roads that wind up and down.

I read on a small sign in a field that a new highway is planned here. It is to cut straight through the hill and the field on top. Red flags indicate the future route. For more than 50 years already, it has been planned to build a highway here, but so far the glacial landscape helped to prevent it. The peat

bogs, which formed between the drumlin hills, are a nature reserve of national importance and the reason why the Federal court rejected the highway project so far. There is a colorful variety of flowers: orchids, Bishop's Wort, and Siberian Iris. Rare dragonflies fly around my head. On a small dam, trains whizz through the idyllic scene.

Power poles are everywhere. They stand there like trees, connected by long threads. In the distance I hear the highway roaring, monotonous, like the buzzing of many bees. The drumlin landscape ends in a huge highway roundabout. Highways from all directions meet and circle in a large radius.

Going downhill, I reach the main valley of the Linth Glacier, on my way to Lake Zurich. Villas and many terraced houses nestle into the topography and collect the sunlight.

A woman asks me what I am doing here and I tell her about my search for the glacier. She proudly explains that she has an erratic boulder in her garden, a red Verrucano from the Glarus region.

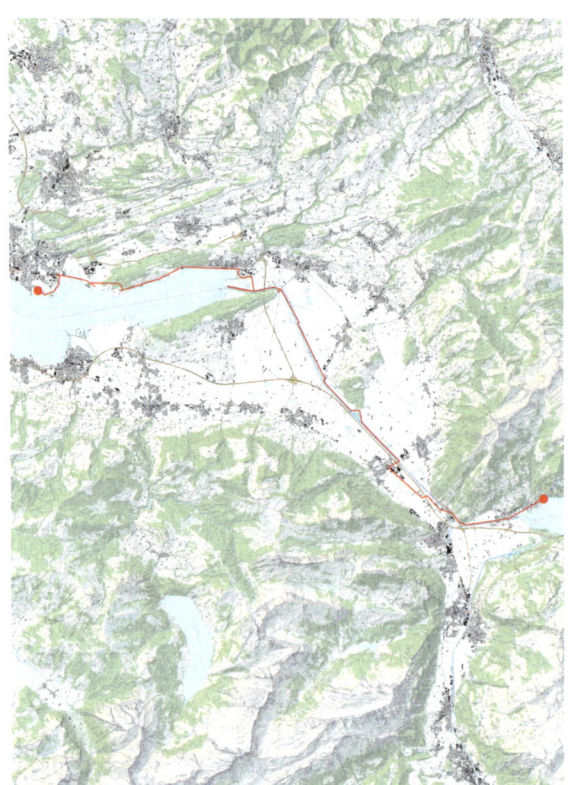

Day 4: Thursday, June 10, 2021, 37.4 km, 10h 5min

I wake up with a view of Lake Zurich. The sun is shining, a bright blue sky. My sunburnt arms hurt a bit. I see the mountains in the distance. They are spotted black and white, like the cows in the fields. Just next to the lake two building sites, a concrete mixer, and a crane. Bungalows with direct access to the lake are being built. Then the hill, covered with terraced houses. Topography and typology. A small airport also uses the flat alluvial land on the other side of the lake. Sport planes take off with loud engines in the direction of the Alps.

One can easily imagine that the lake used to be much bigger, the flat alluvial plain is almost a continuation of the water surface. Around 16,000 years ago, when the glacier just retreated, the water reached all the valleys, a single, huge lake connecting Lake Zurich, Lake Walen, and Lake

Constance. Over the years the rivers filled up most of it with a massive amount of stones and gravel transported from the mountains, from the glacier that was once here.

A dead straight canal divides the plain behind the lake. It is the Linth Canal, famous for being one of the first big infrastructural projects of Switzerland, started already in the eighteenth century. At the river mouth, the water flows fast into the Lake. A couple is swimming, some others came on their bikes to enjoy the view. Looking back towards Zurich, I can picture the glacier between the hills.

I follow the canal for many kilometers, with only a few bends. Alongside electricity

pylons, also dead straight, still accompanying me. The mountains are big now. Small but heavy-looking concrete bunkers appear at regular intervals next to the Linth. They all wear a green hat of moss and grass. Remnants of the Second World War, when

the Swiss Army planned to recreate the huge post-ice-age lake and flood the whole plain in case of attack. On the opposite side of the canal, a concrete factory, also camouflaged in green. Hills of gravel, waiting to be stone again.

Straight ahead, straight ahead. My heels hurt. And my eyes burn. Above my head a glider, silently circling. The main canal is elevated, the side canals are lower. The highway now runs parallel to the canal.

Power poles, power poles, side canal, main canal, side canal, power poles, power poles, highway. Everything is straight, everything is moving, but I am getting tired.

The fact that everything is flat is being used by a farmer to express his opinion on the next vote. He has placed a sign on a retractable forklift. It can probably be seen from the highway.

I discover bunkers everywhere now. Concrete blocks in the meadow. Overgrown, shrouded in green. Better hidden than ever. They have become stones. Like the erratic boulders brought by the glacier.

During the night, the sky is clear. I see many stars, the lights of the cities have dis-appeared.

Day 5: Friday, June 11, 2021, 41.9 km, 10h 32min

Around 1800, the shores of Walensee must have been a miserable place to live. Floods, swamps, and diseases. The water level of the lake was constantly rising as the outflowing river was filling up with stones and gravel. Today, the shore lies peacefully in the sun and no wild river is wriggling through the valley anymore. The canalization solved these problems. Conrad Escher, its engineer, celebrated as a hero, was later called "von der Linth." A title of nobility in gratitude for the canal.

I pass a short piece that was recently renaturalized, gravel beds have formed, small willows are growing. The water is really milky. It comes directly from the glacier. But for a short stretch, the river is allowed to take up some space again. A sign next to it informs me about the little ringed plover, a rare, small bird, which breeds in the reconstructed habitat.

The valley sides are rocky and steep, only few trees find a foothold. It goes almost straight up, over 1,000 m. The bottom of the valley is flat though. I remember the term *U-shaped valley* that I once learnt in school. Steep sides and flat bottoms, a sign of glacial origin. An airport appears on the valley floor, the runway gleaming black from afar. Another airport enabled by the glacier. I follow the straight path along the runway. It is as long as the mountains are high.

I pass a huge erratic boulder, now a real one, not one of the many concrete imitations I came across yesterday. It has a big inscription informing me that the stone has been protected since 1908, after it should have been blown up twice to obtain building material. It is a red Verrucano, like the boulder in the woman's garden.

The valley had been heavily industrialized from the mid-eighteenth century onwards.

I pass by time-honored factories and generous villas of the factory owners, hydroelectric power stations and small canals diverting water. But modern production facilities can still be found, too. The power of the water, of the glacier, lives on and has been transformed.

In the Netstal lime factory, everything is white, even the factory buildings and silos. The limestone has been removed over the years, layer by layer. It has reshaped the mountain, drawn contour lines, small trees have grown on it.

A pharmaceutical factory still produces painkillers, ever since 1949. My feet hurt, too. I continue deliriously along the road.

I see the glacier! Saturday 07.20 p.m., today already 39.6 km walking, and finally it is there! I reach Linthal dead tired. Here I will rest until tomorrow. The destination is now in sight.

Day 6: Saturday, June 12, 2021, 23.8 km, 7h 10min

Two friends join me on the last stage. Finally, I am not walking alone anymore. We continue along the river. Soon a green, geometrical ridge appears. A reservoir is hidden behind the embankment. It is an equalizing reservoir of the Linth-Limmern Hydroelectric Power Plant, Switzerland's largest pumped-storage power plant. The origin of some of the electricity running in the power poles that have accompanied me since the first day. The turbines inside the building behind us make a monotonous, high-pitched sound, a constant hum.

The river looks natural, but there are signs everywhere warning of a sudden rise in water. Large boulders lie in the stream, surrounded by white eddies. Sometimes, fallen trees in between. The stones are much more angular than further down the valley. On the slopes, grey aisles full of debris cut through the forest, at times apparently carrying huge amounts of water.

A bit further, another equalizing reservoir, and another turbine with the same humming, that can be heard from hundreds of meters away. A cable car leads steeply up the mountain to the larger reservoirs behind the massive rock faces. Lake Limmern, built in 1963, and newly constructed Lake Mutt. Traces of the power stations everywhere. The mountain, a single power station.

We continue up an absurdly steep road, only the Alpine taxi and the cars of the power-plant company are allowed to drive up here. The mighty Tödi appears in front of us. A glacier shield at the top. In the middle ages, the local people thought this was the highest mountain in the world. After walking 200 km, I can well understand it. The world is big.

The path zigzags up the slope. You can see that the snow has only recently melted. The

blades of grass still stick flat to the ground. We walk on a moraine overgrown with some grass. The glacier was here only recently, maybe as late as 50 years ago.

Through snowfields up to a plateau, the glacier gets visible in its full size. We slowly approach its edge, the ice that is left. Big crevasses, boulders, and black sand cover its lower part. Still overwhelming. Creeping down from the mountain.

I sit and listen to the sounds of the place. Water rushing, a stream is flowing out of the glacier. In between something like white noise, stones constantly falling. I feel that the glacier is still moving. A big and slow animal. I think of the airports, the lakes, the hills, the power poles, the factories, the valleys, the stones, and the cities. Everything connected to this movement, everything covered with ice. How many days did it snow for it to form, how many days did the sun shine for it to disappear again?

Every step I took was taken by the glacier as well. 6 days and 25,000 years. I fall asleep in a small hut beside it. We are both tired, both at the end of a long journey.

An angular and spiky stone, broken off the mountain, found just next to the glacier. It will be shaped over time, it will move through the land.

Common Water – The Alps

Günther Vogt, Amalia Bonsack, Thomas Kissling, Andreas Klein,
Max Leiß, Roland Charles Shaw, and Rolf Weingartner
(Chair of Günther Vogt, Department of Architecture, ETH Zurich),
Central Pavilion, Giardini della Biennale di Venezia

Photos by Alessandra Chemollo

The Alpine space is much larger than a cursory glance at the relief would suggest. If we look at the water courses, this becomes quite evident. Water is indeed without form, smell, color, or taste. Yet, the element by no means lacks specific properties. In fact, it is the basis of all life.

The Alps have always been of utmost importance as a "water tower" ensuring the very existence of the surrounding landscapes and cities. As a result of today's rising temperatures, however, it is not only the melting of the glaciers that is, quite inexorably, progressing. The gradual dwindling of precipitation in the form of snow also causes the water balance to liquefy. As a result, the Alps can no longer entirely fulfil their function as Central Europe's water tower, especially in summer. However, they remain an important water supply for Central Europe.

This contribution to the exhibition puts the complex problem into a simple format. However, in order to understand the dependencies and interactions between precipitation, temperature, topography, etc. while, at the same time, focusing on the essentials, the respective information cannot simply be conveyed by means of a map or picture. Complementary forms of communication (from videos to photographs to physical models) are needed, which also reflect the approach taken in developing the contribution.

In a broad discourse between the disciplines of natural sciences, engineering, landscape architecture, and art, a proposal for a new reading of the Alps as an "ecological island" was developed. With a view to a larger reference space, i.e., Central and Southern Europe, not only existing uses are gaining in importance in the Alps, but new uses are emerging as well. Among them all, water plays an important role within the Alpine regions as a connecting element as – at least for now – an almost inexhaustible and qualitatively impeccable resource, as a supplier of energy and an attractive feature of landscape with a high ecological and touristic potential. The Alps as a hotspot of biodiversity, their favorable climatic conditions (summer-health resort) or their potential to create new forms of (urban) coexistences – all this and more constitutes the uniqueness of the Alps. In this context, the principles of interaction are crucial. The same as former agrarian societies in the Alpine region, actions should be guided by the principle of a sustainable use of resources. However, a responsible and careful use of the Alpine landscape cannot be guided by traditional ideas. New images and concepts shall have to be developed.

Formless, odorless, colorless, tasteless – boundless? As a civil society, we shall have to come to an understanding about the meaning and value of the landscape. In doing so, a leap in scale shall have to be made. And this not only in terms of space, but also in terms of time. Alpine water in its various states vividly demonstrates this: from the immense volumes of water stored in the form of snow or ice for only a short time or for centuries, to water vapor stored in the atmosphere for a few days, to the streams and rivers that cover the long distances between mountains and valleys for hours or days or even several years (if interspersed with lakes), the water element permeates the manifold dimensions of the overall "landscape system."

Over the Garden Fence

Trachycarpus fortunei – from Ornamental Exotic to Invasive Neophyte

Simon Kroll and Sophie von Einsiedel

A tourist's gaze guided the field trip to Lugano and determined not only the route but also the very medium of documentation. Snapshots of this journey south contrast with historical photographs, postcards, and advertising posters. The palm, once upon a time an attraction and singular phenomenon now is ubiquitous. The exquisite has become a matter of course, and the originally individual a determining aspect.

The native and introduced range of *Trachycarpus fortunei,* as per Gian-Reto Walther et al., "Palms Tracking Climate Change," *Global Ecology and Biogeography* 16 (6), 2007.

> Scientific illustration of *Chamaerops fortunei,* now called *Trachycarpus fortunei,* pictured in the book *Flore des serres et des jardins de l'Europe* by Louis van Houtte, published in 1877.

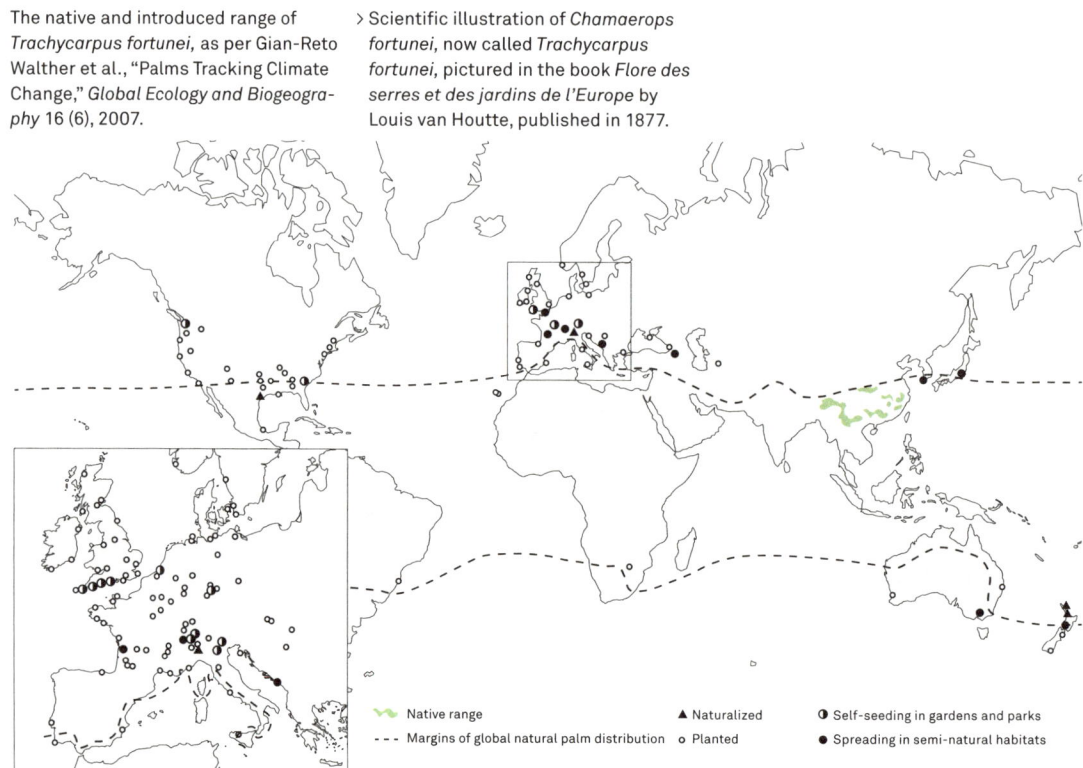

Native range · · · · Margins of global natural palm distribution ▲ Naturalized ○ Planted ◖ Self-seeding in gardens and parks ● Spreading in semi-natural habitats

CHAMÆROPS FORTUNEI *Hook*

♃ *Chine*   *presque rustique.*

A rare exotic brought back as a souvenir from colonial expeditions to China, *Trachycarpus fortunei,* became a symbol of sophistication and luxury across Europe throughout the eighteenth and nineteenth centuries. Favored by the expansion of cross-Alpine travel into mass tourism and by a globally warming climate, the Chinese windmill palm has evolved from a cultural icon into an invasive species spreading into Alpine landscapes and crowding out native vegetation and wildlife.

A Warm Welcome

Arriving in Lugano on the other side of the Bernese Alps after the depths of the Gotthard tunnel is like entering another world. Immediately, the palm trees gently swaying in the wind let us know that we have arrived in the south. Yet, this ubiquitous plant is itself a tourist to the region; one who has long outstayed its welcome though. Originally native to the mountain forests of southern China, *Trachycarpus fortunei,* commonly known as the Chinese windmill palm, was introduced to the Canton of Ticino as an ornamental plant in the eighteenth century.[1] *Trachycarpus fortunei*, named after British plant hunter Robert Fortune, came to Europe as a souvenir of expeditions to the British colonies in Asia. Displayed alongside other "exotics" in glasshouses such as the Palm House in Kew Gardens, it symbolized the power and domination of the British Empire, but also the perceived sophistication and superiority of

British and – by association – European culture and society.[2] Just as orangeries in previous centuries, during the Victorian age, palm houses and their exotic inhabitants were costly to maintain and only a small elite was able to afford them.[3] This exclusivity transformed palm trees from specimens of botanic novelty into icons of luxury that spread across Europe. Soon palm trees were found in the lobby during winter and driveway in summer of every luxury hotel to make their upper-class residents feel at home as they travelled through Europe on their Grand Tour.[4] Where the climate allowed, palm trees were planted along promenades and in public parks to stage idealized landscapes of an imagined paradise in the South.

Souvenirs of the South

The onset of rail travel in the late nineteenth century and the completion of the Gotthard tunnel through the Alps in 1882 made travel more affordable; travelers became tourists.[5] As they visited Lugano, sent postcards, and bought panorama pictures as souvenirs of their travels, they distributed images of

Lugano's hotels, villas, parks, squares, and their palm trees throughout Europe. This made *Trachycarpus fortunei* into a defining feature of the imagined – and later actual – landscapes of Lugano, and the South of the Alps. In the early twentieth century, tourism advertising picked up on the palm trees as a symbol of luxurious travel that was now becoming accessible to a wider population. Today, posters, postcards, and holiday snapshots of places in and around Lugano still show a palm tree in front of the lake and mountains in the background.

A Piece of Paradise
Far beyond the lakeside of Lake Lugano, the Chinese windmill palm came to represent a promise of the idyllic landscapes of the South and the idle luxury associated with these locations. Already a fixture in the canon of popular house plants in European homes since the late nineteenth century, *Trachycarpus fortunei* alluded not only to the far-away exotic, but soon also to more familiar visions of sunny summer holidays south of the Alps. With the expansion of plant nurseries and the birth of garden centers post WWII,[6] the plant itself became the souvenir. As one of the most sold plants in Switzerland, it can be bought in every garden center and even at supermarkets. Available in all shapes and sizes, *Trachycarpus fortunei* is found in every private garden, backyard, and on every balcony. After a tumultuous journey from southeast China to Britain and Lugano, the Chinese windmill

palm is now firmly established in the garden landscapes of Ticino and beyond.[7] A small piece of paradise accessible to all.

Jumping the Fence
But the resilient and wide-seeding palm did not content itself with its prominent residence in nearly all parks and gardens around Lake Lugano, Lake Geneva, and Lake Lucerne. The lake landscapes around the Alps have a specific microclimate, in which the lake reduces temperature ranges and the reflection of sunrays on the water produces high levels of insolation on adjacent hills.[8] Aided by these favorable conditions and

increasing summer temperatures due to global warming, the palm jumped the garden fence and made its way into nearby woodlands and forests.[9] The reduced intensity of forest management since the 1950s allowed *Trachycarpus fortunei* to establish itself in the undergrowth, rapidly becoming one of the most invasive species in Switzerland and crowding out local biodiversity.[10] Attempts to manually eradicate the plant have so far been futile; the dream of exotic luxury under the Mediterranean sun is fast becoming an ecological nightmare.

A Day Trip Back in Time
Travelling to Ticino like a tourist, we traced the route of *Trachycarpus fortunei* through space and time in the garden landscapes of Lugano, mapping its path from a rare exotic

5174 Palme a Castagnola col Monte S. Salvatore e Lugano

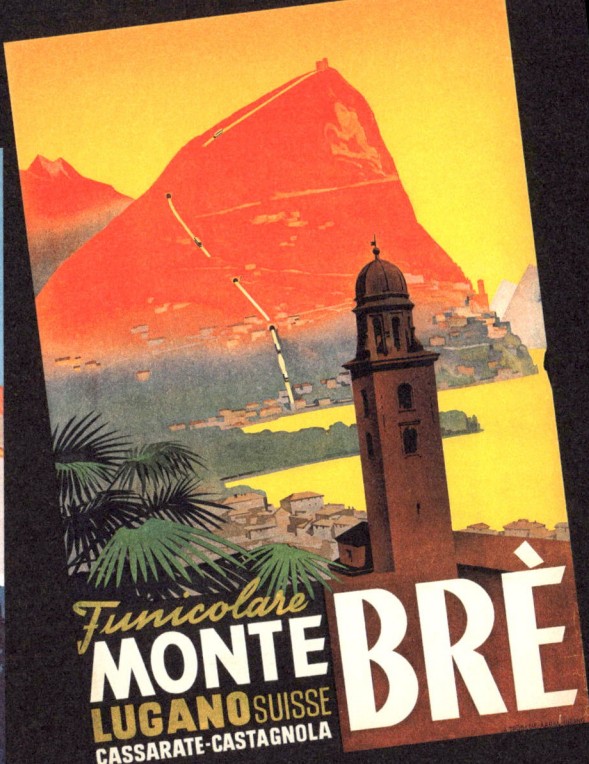

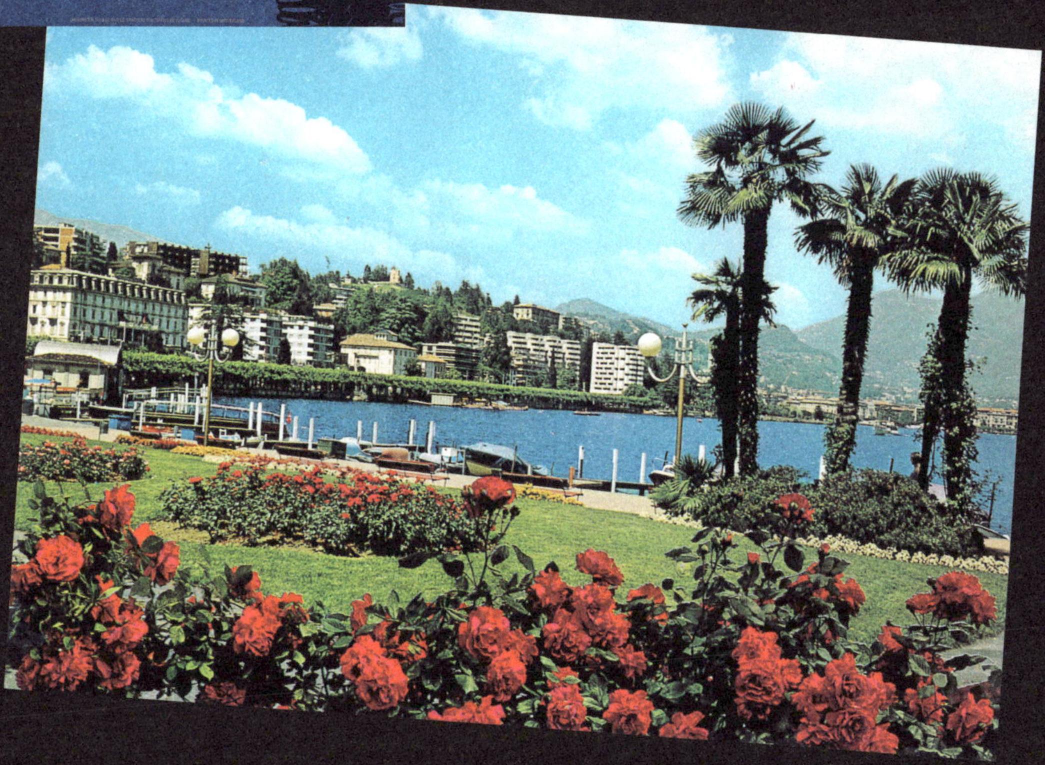

181 Lugano-Paradiso, Quai

to a ubiquitous invader through a selection of souvenirs.

Starting out in Gandria, once a small fishing village turned into a quaint day-trip destination, we tracked the spread of the Chinese windmill palm along the lake shores to downtown Lugano. The narrow, stony lanes of Gandria only harbored one or two specimens of *Trachycarpus fortunei*, but as soon as we entered the open landscape, the palms were ubiquitous. Once planted along the water front of secluded hotels and villas, they have colonized the surrounding hills to well over 50m above lake level.

Closer to Lugano, prominent lakeside villas such as the historic Villa Heleneum and Villa Favorita all displayed fully grown palms along the driveway and waterfront. More contemporary villas built further up the hill picked up on the motif of the palm tree in greater variation, from individual plants framing entrances and views to dense stands of palms in front yards and gardens, to small pot plants on the balconies of apartment buildings.

In Lugano itself, we observed the use of *Trachycarpus fortunei* in the public spaces of the city. Dense groups of palms in central Parco Ciani contrasted with straight lines of tall palms along the public swimming pool and the historic boulevard and waterfront of Paradiso, an elegant neighborhood of Lugano. In Paradiso, mature palm trees surrounding the remnants of nineteenth-century grand hotels, such as Hotel Splendide or Hotel Victoria, pointed towards the origins of the association between the palm tree and Ticino. But while the grandeur of these hotels may have faded, the palm trees remained an icon of luxury that is appropriated even by modern hotels such as Hotel Eden in all its 1980s boldness, featuring *Trachycarpus fortunei* in great abundance.

After a short drive, we reached Parco Scherrer, an elaborately designed private garden from the 1930s, where the palm was employed as a key design feature to evoke the "exotic" and "foreign" alongside Arabian and East Asian pavilions dotted around the estate.

Returning to the train station, we bought souvenirs – like any real tourist would – at a local kiosk. Nearly all postcards and travel posters depicted an idealized Ticino landscape with *Trachycarpus fortunei* in front of Monte Brè. However, the best find was no doubt a small colorful snow globe, containing an idyllic landscape of a traditional stone house, a cow, and two large palm trees. No

souvenir could have better summarized the persistent presence of *Trachycarpus fortunei* in the imagined landscapes of Ticino. Judging from the palms rapid colonization of the Alpine forests supported by a warming climate, this landscape of palm trees and cows in the snow – labelled by the snow globe as "Switzerland" – might soon become reality.

1 Gian-Reto Walther, *Laurophyllisation in Switzerland*, Ph. D. diss., Swiss Federal Institute of Technology (ETH), 2000. https://doi.org/10.3929/ethz-a-003886833.
2 Carolyn Fry, *The Plant Hunters*, André Deutsch Ltd, London, 2009.
3 Heinrich Hamann, *Goldorangen, Lorbeer und Palmen – Orangeriekultur vom 16. bis 19. Jahrhundert*, Michael Imhof Verlag, Petersburg, 2010.
4 Günther Vogt et al., *Mutation and Morphosis: Landscape as Aggregate*, Lars Müller Publishers, Zurich, 2020.
5 Roland Flückiger-Seiler, *Hotelpaläste des Historismus in der Schweiz*, Historische Hotels Schweiz, 2010.
6 BBC Local, *Garden Centres "First Began In Ferndown, Dorset,"* BBC Local, 2010.
7 Vincent Fehr, *Aspekte der Laurophyllisierung der Wälder im Gebiet Südtessin (CH) – Lago Maggiore (IT) und die globale Erwärmung*, University of Zurich, Department of Geography, 2014.
8 See note 4 above.
9 Gian-Reto Walther et al., "Palms Tracking Climate Change," *Global Ecology and Biogeography* 16 (6), 2007, pp. 801–809. https://doi.org/10.1111/j.1466-8238.2007.00328.x.
10 See note 7 above.

Late-Glacial and Post-Glacial Immigration and Dispersal of Important Tree Species in Switzerland

Conradin A. Burga

Data Basis, Overview of the Forest Development in the Late-Glacial and the Post-Glacial Periods

The history of the Late-Glacial and Post-Glacial immigration and dispersal of important tree species in Switzerland presented here is based on the first Swiss database of pollen analyses covering the last 250,000 years, compiled by the author from 1989 till 1994. For this purpose, a total of 618 peat bogs and lake sites in Switzerland were subjected to pollen analyses, including a total of 211 localities with absolutely dated borehole profiles (radiocarbon ages) (Fig. 1).

The immigration and dispersal of the different plant species were presented for the different time periods in the form of point-and-time isolines. In addition, the temporal-spatial dispersal of plant species or genera was recorded in altitude-time diagrams for the last 250,000 years, according to the biogeographical division of Switzerland into the regions of Jura, Central Plateau, Pre-Alps, Central Alps, and Southern Alps.[1]

Vegetation dynamics during the Late-Glacial and Early Holocene mainly reflect natural environmental processes. About 15,000 years ago, after the Late-Glacial, Alpine glaciers melted more quickly and the first conifers such as Scots pine (*Pinus sylvestris*), mountain pine (*Pinus montana*), Swiss stone pine (*Pinus cembra*), European larch (*Larix decidua*), and silver birch (*Betula pendula*) returned during the warmer periods (Bølling and Allerød interstadials) from their glacial refugia. Between 13,000 and 12,000 years ago, the oldest coniferous forest communities in Switzerland were formed with the European

1 Conradin A. Burga, Roger Perret, *Vegetation und Klima der Schweiz seit dem jüngeren Eiszeitalter*, Ott Verlag, Thun, 1998.

Fig. 1: Swiss sites with absolutely dated borehole profiles (radiocarbon age of peat bogs and lake boreholes). Conradin A. Burga & Roger Perret, *Vegetation und Klima der Schweiz seit dem jüngeren Eiszeitalter*, Ott Verlag, Thun, 1998.

larch-Swiss stone pine woods, and mountain-pine forests. The Last Glacial (LG) ended with the last rebound of the Younger Dryas cold phase, which lasted about 1,000 years. Around 11,000 years before present (years BP), the Holocene or Post-Glacial period began, during which global temperature rapidly rose for about a millennium to almost today's values, thus enabling the immigration and dispersal of warmth-loving trees and shrubs such as oak (*Quercus* spec.), lime (*Tilia* spec.), elm (*Ulmus* spec.), maple (*Acer* spec.), ash (*Fraxinus excelsior*), Norway spruce (*Picea abies*), silver fir (*Abies alba*), beech (*Fagus sylvatica*), hazel (*Corylus avellana*), and alder (*Alnus* spec.).[2]

As a result of human land appropriation, an increasing human influence on the vegetation and landscape began at the transition from the Mesolithic to the Neolithic (Mesolithic/New Stone Age) around ca. 7,000 years BP. From this time onwards, there is evidence of sparser, cleared forests in the lowlands and, somewhat later in the Alps, of the lowering of timberlines to obtain arable land, Alpine meadows, and pastures with the help of pollen analyses. During the last 10,000 years, the natural Alpine timberline oscillated up and down by about 100 meters due to various cold phases.[3] Due to the increasing influence of land use and settlement, the Alpine timberline has, especially since the Middle Ages, been lowered by several hundred meters in altitude, depending on the region. The immigration and dispersal of plants show a complex interplay of highly different biotic and abiotic

2 Conradin A. Burga, "Holozäne Klimaänderungen und Waldgrenzschwankungen in den Alpen," in *Warnsignal Klima. Hochgebirge im Wandel. Wissenschaftliche Auswertungen,* ed. José L. Lozan et al., GEO, Hamburg, 2020.
3 See note 2.

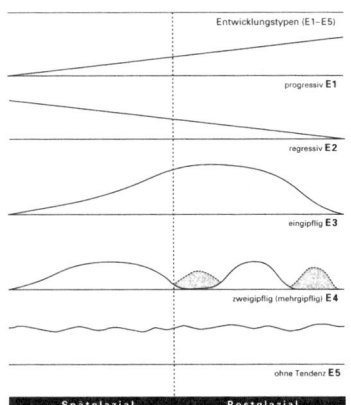

Fig. 2: Late-Glacial and Post-Glacial immigration and dispersal patterns of plants. Conradin A. Burga, "Late- and post-glacial immigration patterns of some important plant species in Switzerland," *Écologie* 29 (1–2), 1998, pp. 233–237.

factors. These include dispersal biology and species-specific plant characteristics, competition with other plants (e.g. root and light competition), and influences by mammals (browsing, grazing) and pests (insects, fungi). Site factors (topography, soil conditions, available water, microclimate, and local climate) are the decisive factors of any colonization and succession processes of plant stands. With regard to the topic dealt with here, the distance to glacial refugia, glacial survival sites on nunataks, and migration routes play the main role.

In general, the following patterns of the immigration and dispersal of plants since the Late-Glacial can be distinguished: 1. progressive development (E1), 2. regressive development (e.g. of glacial floras) (E2), 3. dispersal development with a maximum (E3), 4. dispersal development in several phases with several maxima (E4), and 5. immigration and dispersal without a discernible trend or with little information available as yet (E5) (Fig. 2).

The Late-Glacial Immigration and Dispersal of the Swiss Stone Pine

As a raw-humus seedling, the Swiss stone pine (*Pinus cembra*), a typical conifer of the continental high valleys of the Central Alps of Valais and the Grisons, grows mainly on podzols and, in contrast to the European larch, is not a typical pioneer tree. The frost hardiness of the Swiss stone pine amounts to −47°C, that of the European larch to −40°C. Today's European Swiss stone-pine area consists of two separate sub-areas: (1) the high valleys of the Central Alps (Argentera-Mercantour in Italy/France, and Valais and Engadine in Switzerland) and (2) the Carpathians (High Tatras).

Glacial Swiss stone pine refugia are to be found on the southern rim of the Alps and in the south-eastern Alpine region as well as in the Carpathians, with the first two areas most likely particularly important for the Late-Glacial immigration into Switzerland. Thus, the glacial refugia were not very far from today's stands, otherwise a rapid Late-Glacial recolonization of the Southern Alps and Eastern Central and Pre-Alps of Switzerland would not have been feasible. Already for the Late-Glacial,

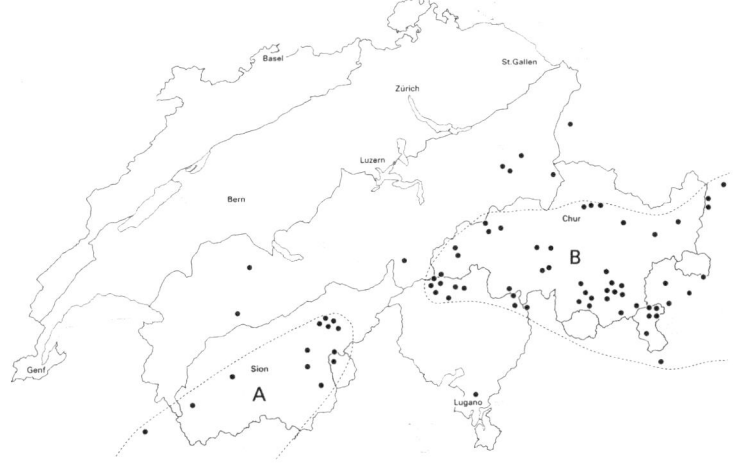

Fig. 3: The two Swiss stone pine sub-areas in the Swiss Alps. A: western area (Valais, Pennine Alps), B: eastern area (North Ticino-Grisons-Churfirsten-Alpstein). Burga & Perret, *Vegetation und Klima*, 1998.

the two present-day sub-areas of Swiss stone pine of the Pennine and Rhaetian Alps can be identified. In the Southern Alps, the Swiss stone pine migrated at about the same time as the European larch (probably originating from the same glacial refugia) already towards the end of the Oldest Dryas (ca. 14,000 years BP) and, in the Allerød (from ca. 12,000 years BP), formed a European larch-Swiss stone pine forest belt up to ca. 1,700 m a.s.l.

Before the Norway spruce immigration, the latter extended from the Gotthard southwards to the Camoghè-Tamaro-Limidario line. In the Grisons, Central Alps, and eastern Pre-Alps, the Swiss stone pine first appeared in the Bølling (from ca. 13,000 years BP). In the area of the Lukmanier Pass and Vorderrhein/Hinterrhein Valleys, a belt of European larch-Swiss stone pine forest formed in the Allerød up to 1,900 m a.s.l. Compared to the eastern Central Alps, the Swiss stone pine appeared later in Valais and further southwest in the Dauphiné (France), indicating an east-west migration.[4] Rikli (1909) attributes this to the unfavorable climate and points to the steep topography as the reason for the two separate Swiss stone pine areas in Valais and Engadine with their tree line at 2,300–2,400 m a.s.l. and a lower dispersal in the Ticino Alps ("Ticino gap"). However, Rikli and Hager[5] were able to infer an earlier and larger dispersal based on Swiss stone pine fossil finds (nut fruits, wood) and field names in areas that were free of Swiss stone pines at the time.

4 Martin Rikli, *Die Arve in der Schweiz*, Denkschr. Schweiz. Naturf. Ges. 44, Georg & Cie., Basel, 1909.
5 Karl Hager, *Verbreitung der wildwachsenden Holzarten im Vorderrheintal (Kt. Graubünden)*, Erhebungen über die Verbreitung der wildwachsenden Holzarten in der Schweiz, Lieferung 3, Bern, 1916.

Fig. 4: Post-Glacial Norway spruce immigration into Switzerland between 8,000 and 5,000 years BP, generally from east to west. Burga & Perret, *Vegetation und Klima*, 1998.

Basel
St. Gallen
Zürich
Luzern
Bern
Chur
Sion
Genf
Lugano
5500
6500
8000

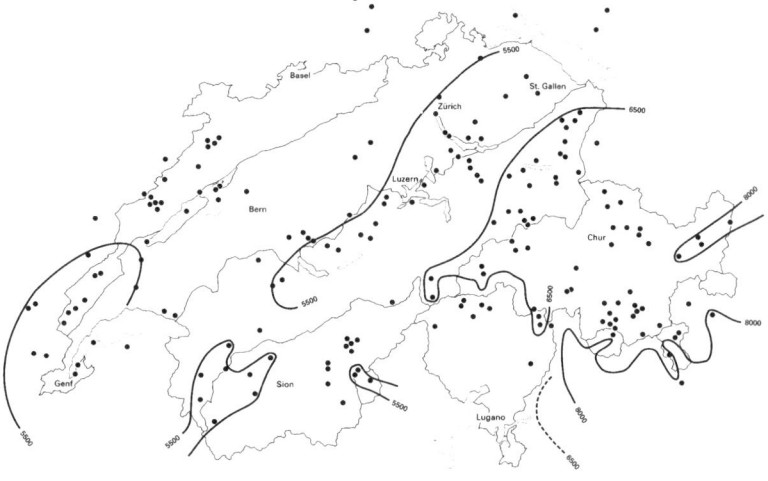

Fig. 5: Migration rates of Norway spruce in the Rhine valley and across the San Bernardino pass area. (Original Conradin A. Burga unpubl.).

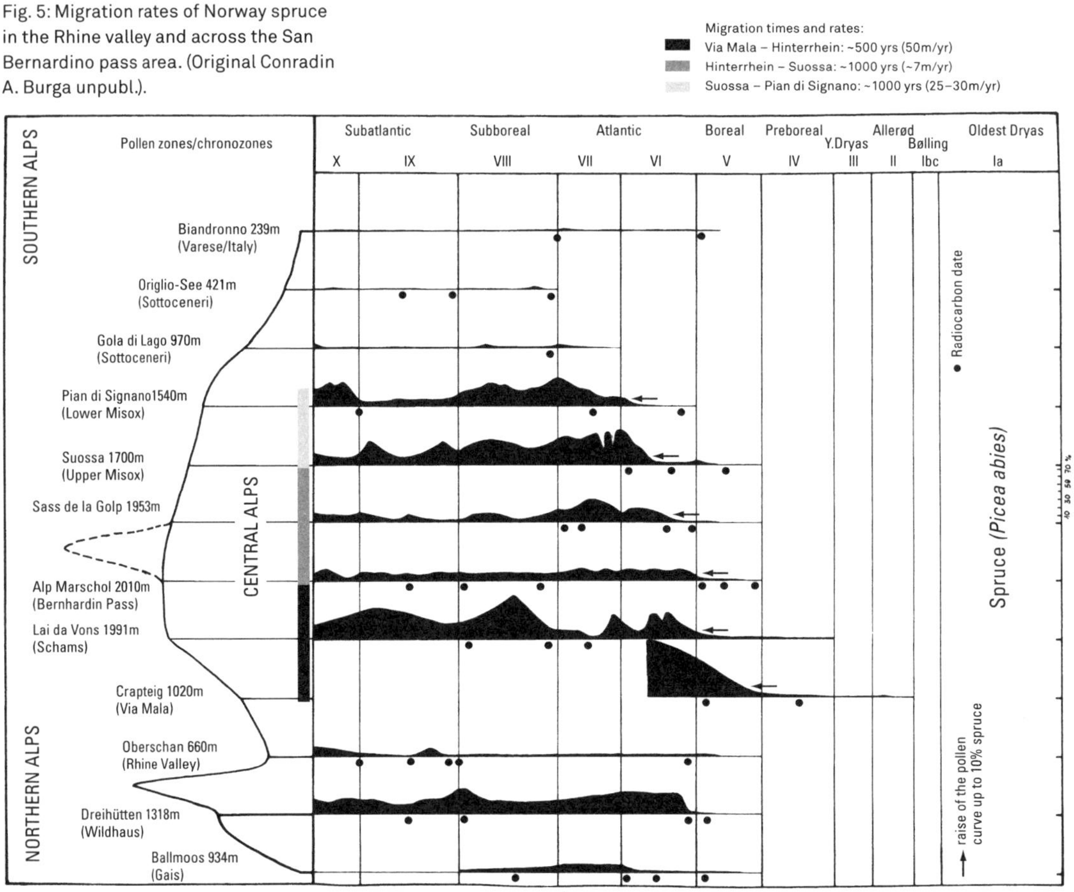

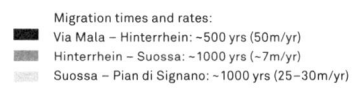

Migration times and rates:
Via Mala – Hinterrhein: ~500 yrs (50m/yr)
Hinterrhein – Suossa: ~1000 yrs (~7m/yr)
Suossa – Pian di Signano: ~1000 yrs (25–30m/yr)
SOUTHERN ALPS
CENTRAL ALPS
NORTHERN ALPS
Pollen zones/chronozones
Subatlantic
Subboreal
Atlantic
Boreal
Preboreal
Y.Dryas
Allerød
Bølling
Oldest Dryas
X
IX
VIII
VII
VI
V
IV
III
II
Ibc
Ia
Biandronno 239m (Varese/Italy)
Origlio-See 421m (Sottoceneri)
Gola di Lago 970m (Sottoceneri)
Pian di Signano1540m (Lower Misox)
Suossa 1700m (Upper Misox)
Sass de la Golp 1953m
Alp Marschol 2010m (Bernhardin Pass)
Lai da Vons 1991m (Schams)
Crapteig 1020m (Via Mala)
Oberschan 660m (Rhine Valley)
Dreihütten 1318m (Wildhaus)
Ballmoos 934m (Gais)
● Radiocarbon date
Spruce (Picea abies)
← raise of the pollen curve up to 10% spruce

Post-Glacial Immigration and Dispersal of Norway Spruce from Southern Europe

Today, Norway spruce (*Picea abies*) grows to the following upper altitudinal limits in the Swiss Alps: stand limit ca. 1,700–2,000 m, tree line ca. 1,900–2,200 m, stunted-tree line ca. 2,000–2,400 m a.s.l. The stand limit coincides more or less with the 10°C July isotherm and is higher in continental areas (Central Alps) than in the oceanic locations of the Pre-Alps and Southern Alps. However, winter frost resistance is very high at −40°C. Today, the Eurasian conifer is the Norway spruce, consisting of the two subspecies, *Picea abies* ssp. *abies* (European spruce) and *P. abies* ssp. *obovata* (Siberian spruce), and forms three sub-areas: a) Central and South-Eastern Europe (Alps, Jura, Black Forest, Thuringia, Harz, Bavarian Forest, Bohemian Forest, Carpathians, and Balkans), b) North-Eastern Europe (Scandinavia, Baltic States, Russia, and Urals), and c) Siberia (Urals-Okhotsk Sea). Norway spruce forms pure stands in the subalpine blueberry-spruce forests of the Alps and also mixed stands with silver firs, beeches, European larches, and pines. The natural Norway spruce stands in Switzerland are distributed among 17 forest communities.

In Europe, four glacial refugia of Norway spruce from the Last Glacial (LG, "Würm Glacial") are assumed: a) Western and Central Russia, b) the Carpathian Mountains of Poland and Romania, c) the Balkan Peninsula, especially the Dinaric Mountains, and d) probably the eastern edge of the Alps in Austria and Slovenia.[6] Probably starting from refugia on the eastern edge of the Alps or in the Dinaric Mountains, Norway spruce migrated into the Upper Adige Valley already around 9,000 years BP. It then crossed the Tonale and Aprica Passes along the southern Alpine rim towards the west and spread to the Valtellina around 8,000 years BP; from here, it reached the Val Poschiavo and the Reschen Pass via the Val d'Adige.

Through the Inn valley, Norway spruce reached Lower Engadine and around 8,000–7,500 years BP spread further into Upper Engadine.[7] Further west, around 7,500–6,500 years BP, Norway spruce was able to advance from the Vorarlberg and through the Rhine Valley southwards

6 See note 1.
7 Conradin A. Burga, *Gletscher- und Vegetationsgeschichte der Südrätischen Alpen seit der Späteiszeit*, Denkschr. Schweiz. Naturf. Ges. 101, Verlag Birkhäuser, Basel 1987. See also note 1.

Fig. 6: Post-Glacial silver fir immigration into Switzerland in the time between 9,000–6,000 years BP, generally from south to north. Burga & Perret, *Vegetation und Klima*, 1998.

into the Vorderrhein and Hinterrhein Valleys, the Oberhalbstein and over the San Bernardino Pass into Misox down to northern Ticino.[8] (Fig. 4) The N-S migration profile of Norway spruce from the St. Gall Rhine Valley over the San Bernardino Pass into Ticino and to the southern edge of the Alps (Italy) provides rare information about its migration speed. (Fig. 5) For the Via Mala-Hinterrhein section, Norway spruce needed about 500 years (migration rate 50 m/year); for the Hinterrhein-Suossa/ San Bernardino-Village section, however, about 1,000 years (only about 7 m/year). This was due to the climatically demanding crossing of the pass and the competition from mountain pines, green alder shrubs, and dwarf shrub heath. For the Suossa-Pian di Signano/ Grono section in the Lower Misox, the Norway spruce needed another ca. 1,000 years (25–30 m/year). This slow migration was due to the strong competition of deciduous trees and shrubs and silver firs on the southern side of the Alps. The current, natural Norway spruce belt tapers out at Monte Laura and Monte d'Arbino east of Bellinzona.

An average migration rate of 60–500 m/year is given for Norway spruce. The migration rates of other tree species are: silver fir 40–300 m/year, Scots pine 1500 m/year, beech 175–350 m/year, oak 5–500 m/year, lime 50–500 m/year, elm 100–1000 m/year, and silver birch 250–<2000 m/year.[9] Here, the mode of of pollen and seed dispersal is crucial as most forest trees are anemophilous or windborne; beech and oak seeds are, inter alia, dispersed by

8 Conradin A. Burga, *Pollenanalytische Untersuchungen zur Vegetationsgeschichte des Schams und des San Bernardino-Passgebietes (Graubünden, Schweiz)*, Diss. Bot. 56, Cramer, Vaduz, 1980.
9 Gerhard Lang, *Quartäre Vegetationsgeschichte Europas*, Fischer, Jena, 1994.

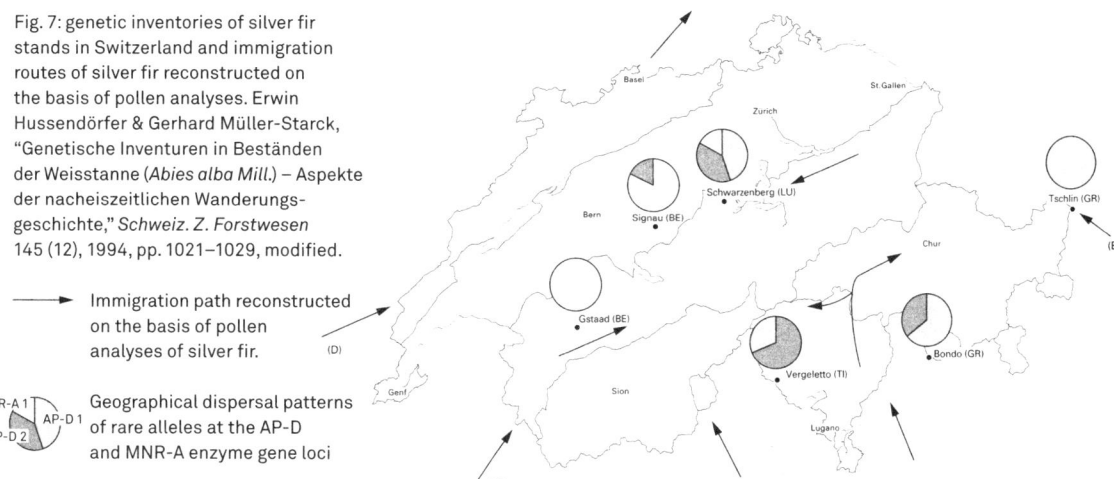

Fig. 7: genetic inventories of silver fir stands in Switzerland and immigration routes of silver fir reconstructed on the basis of pollen analyses. Erwin Hussendörfer & Gerhard Müller-Starck, "Genetische Inventuren in Beständen der Weisstanne (*Abies alba* Mill.) – Aspekte der nacheiszeitlichen Wanderungsgeschichte," *Schweiz. Z. Forstwesen* 145 (12), 1994, pp. 1021–1029, modified.

⟶ Immigration path reconstructed on the basis of pollen analyses of silver fir.

Geographical dispersal patterns of rare alleles at the AP-D and MNR-A enzyme gene loci

animals. Some trees (silver birch, oak, Norway spruce) are also known for "dispersal jumps" of 2 to 15 km.

Around 5,500 years BP, the Norway spruce had spread to the line formed by Lake Constance/Lower Lake Constance-Zurich-Entlebuch-Thun-Simmental. Further west, the Norway spruce had to compete against deciduous trees such as beech and oak and reached its ecological limit of dispersal. At the same time, the Norway spruce reached Upper Valais from the south-east or south-west over the Simplon Pass along the southern Alpine rim and Lower Valais from Savoy over the Col de la Forclaz. It also reached SW Jura around 5,500 years BP via Savoy, which is indicated by the highland spruce type, the so-called Risoux spruce.[10] This separate migration route from Savoy into Lower Valais and SW Jura is confirmed by genetic inventories of recent spruce stands.[11]

The Holocene Immigration and Dispersal of Silver Fir from SE and SW

Today, silver fir (*Abies alba*), a late frost-sensitive shade wood, forms forest communities of the colline and montane altitudinal zones together with Norway spruce and beech. The last glacial silver-fir refugia were located on the southern Balkan peninsula, in Anatolia, and the southern Apennines. In southern Switzerland (Ticino), the silver fir arrived around 9,000 years BP via two branches of immigration from the southeast (Bergamasque, Tridentine, and Venetian Alps) and from the southwest

10 See note 1.
11 Gerhard Müller-Starck, "Genetic Variation in High Elevated Populations of Norway Spruce (Picea abies (L.) KARST) in Switzerland," *Silvae Genetica* 44 (5–6), 1995, pp. 356–362.

Fig. 8: overview of immigration routes of some forest-forming trees into Switzerland. Conradin A. Burga, "Swiss Vegetation history during the last 18,000 years," *New Phytol.* 110, 1988, pp. 581–602.

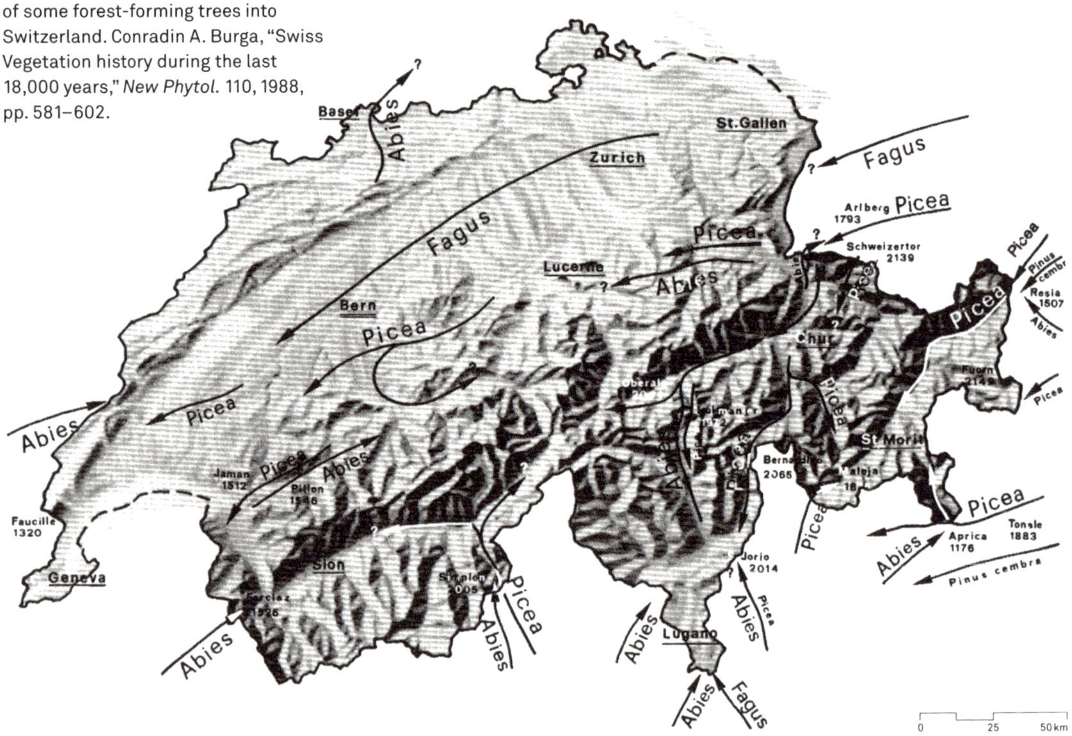

Fig. 9: altitude-time diagram showing the development of the Central Alpine forest belt of the Central Grisons and the course of the forest and snowline since the Late Glacial. Conradin A. Burga, "Holozäne Klimaänderungen und Waldgrenzschwankungen in den Alpen," in: José L. Lozan et al., *Warnsignal Klima. Hochgebirge im Wandel. Wissenschaftliche Auswertungen,* GEO, Hamburg, 2020, pp. 103–108.

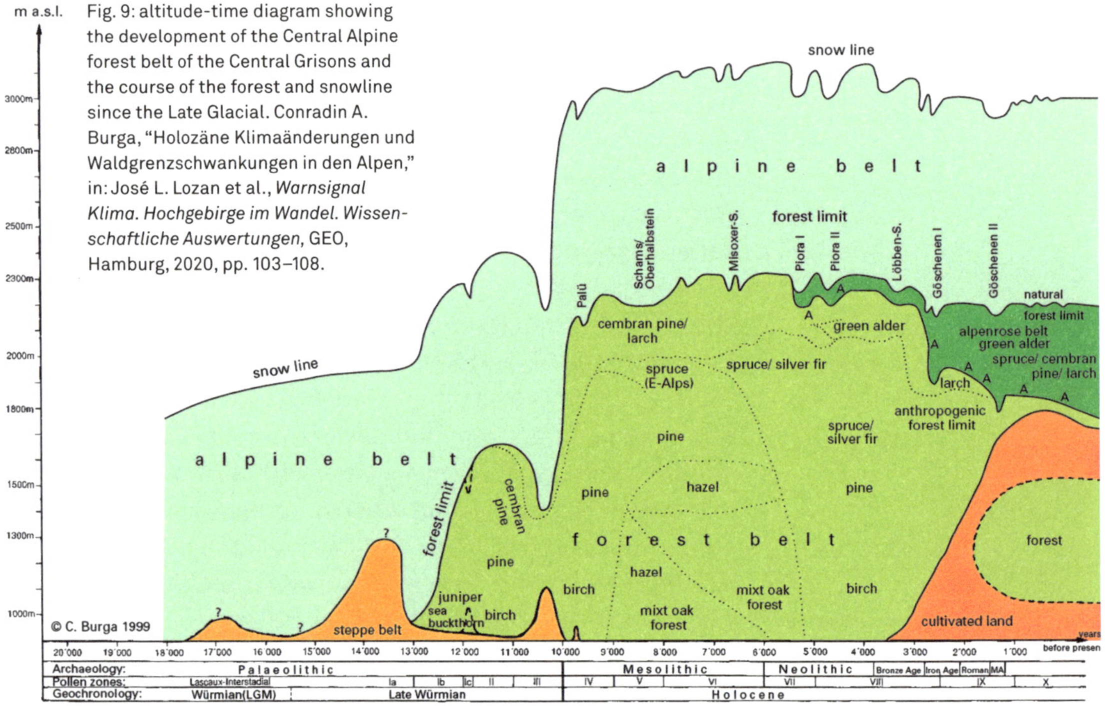

(Ligurian Apennines, Cottian and Piedmontese Alps), after it had already appeared on the southern edge of the Alps in northern Italy about 12,000–10,000 years ago. The rate of migration from its glacial refugia in southern Italy and the Balkans may have been 75–120 m/year.[12] (Fig. 6) From southern Ticino, the silver fir rapidly migrated northwards via Leventina and Misox. The Lukmanier, Simplon, Forclaz, Reschen, and Brenner passes acted as migration corridors into the northern Alps of Switzerland and Austria. About 7,000 years BP, the dispersal front reached the Prättigau-Appenzell-Zurich Oberland-Rigi-Hohgant-Simmental line. Around 6,000 years BP, the silver fir reached the northern Swiss Central Plateau (Lake Constance-Schaffhausen), the western Central Plateau and Central Jura. It probably reached SW Jura along a separate branch of immigration via the Grande Chartreuse and entered the Vallée de Joux shortly before the Norway spruce did. At high altitudes, silver fir-beech forests predominated. Thanks to genetic inventories of recent silver-fir stands in Switzerland, these immigration and dispersal routes could be confirmed.[13] (Fig. 7)

The Immigration Paths and the Altitudinal Dispersal of Forest-Forming Trees and Formation of Today's Vegetation Belts

When looking at the immigration paths of the Norway spruce and silver fir using the timelines in years (Figs. 4 and 6), initially only a two-dimensional picture emerges. It should be noted, however, that the timelines are to be understood as superimposed on a virtual relief, which includes the spatial component. An overview of the Holocene immigration routes of the most important forest-forming tree species in Switzerland with underlying relief is given in Figure 8. Some proven or presumed pass crossings are also listed.

The altitude-time diagram of the development of the Central Alpine forest belt of the Central Grisons since the Late-Glacial shows the formation of the forest belts and the forest-free grass belts over the course of the last 18,000 years (Fig. 9).

12 Gerhard Lang, "Some Aspects of European Late- and Post-glacial Flora History," *Acta Bot. Fenn.* 144, 1992, pp. 1–17.
13 Conradin A. Burga & Erwin Hussendörfer, "Vegetation History of Abies Alba Mill. (Silver Fir) in Switzerland – Pollen Analytical and Genetic Surveys Related to Aspects of Vegetation History of Picea Abies (L.) H. Karsten (Norway Spruce)," *Veget Hist Archaeobot* 10, 2001, pp. 151–159.

Until the beginning of the Late-Glacial reforestation with silver birch, Scots pine, and mountain pine about 13,000 years ago (Oldest Dryas/Bølling interstadial), herbaceous pioneer vegetation with first woody plants such as willow (*Salix* spec.), and on dry sites sea buckthorn (*Hippophae rhamnoides*) and juniper (*Juniperus communis*), comparable to today's newly cleared glacial forelands, prevailed on the areas that became ice-free. In addition, in continental regions on sites unable to support forests, there were sometimes extensive steppes consisting of grasses, wormwood species (*Artemisia* spec.), and other herbs. The major Late-Glacial reforestation thrust occurred about 12,000 years ago (Allerød interstadial), when the Alpine timberline of Swiss stone pine, European larch, and mountain pine rapidly rose to about 1,700 m. Approximately parallel to this, the climatic snowline rose to about 2,500 m a.s.l., i.e., a broad altitudinal layer of alpine herb meadows and scree vegetation established itself between forest and snowline.

As mentioned above, around 11,000 years BP, at the turn of the Holocene, the last Glacial cold rebound of the Younger Dryas period occurred with the renewed and marked advances of the Alpine glaciers. This caused a strong lowering of the timberline by 200–300 meters in altitude. Due to the rapid temperature rise at the beginning of the Holocene, the Alpine timberline rapidly rose in the following Preboreal to altitudes comparable to today's to around 2,300 m a.s.l. Before human influence from the Neolithic onwards, the potentially natural timberline oscillated up and down by an average of 50–180 altitude meters as a result of several cold and warm periods (hypsithermal).[14] In the course of the Holocene, the other forest-forming tree species migrated and dispersed into the ecological niche appropriate to them (altitude level, mountain exposure, soil type, etc.), according to their site requirements and the local competition from other species, as shown above for Swiss stone pine, Norway spruce, and silver fir for the subalpine and montane altitudinal levels. The altitude-time diagram also shows the immigration and dispersal of thermophilic deciduous woods (mainly mixed oak forest) with hazel in the colline and montane belts and

14 See note 2.

the green alder at the timberline (Fig. 9). Between the potential-natural timberline and the climatic snowline, an area of around 800 m altitude interval established itself with herb meadows, scree, and pioneer plant communities of the alpine and lower nival altitude levels. In the course of the Neolithic, the anthropogenic timberline, lowered by several hundred meters in altitude due to alpine pasture clearing and timber extraction, began to form.

On the basis of the pollen profiles and plant fossils, Figs. 10–12 present a reconstruction of the vegetation of Switzerland from the early Holocene (Preboreal, ca. 9,500 years BP) to the Subboreal (5,000–2,500 years BP), showing the emergence and change of deciduous and coniferous forests and the changes from the colline to the nival vegetation belt.

In the early *Preboreal*, birch-pine forests dominated on the Central Plateau and in low areas of the northern and southern Alpine valleys; in southern Ticino, the first stands of mixed oak forest already grew. Dry sites on the valley bottoms and sunny slopes of the Rhone, Rhine, and Inn (or *En* in Romansh) in Lower Engadine were largely occupied by Scots pine and steppe or grassland plains. In the high altitudes of the Central and Southern Alps, European larch-Swiss stone pine forests and extensive subalpine dwarf shrub heaths grew, above which alpine herb meadows, and rock and scree plants flourished (Fig. 10).

About 1,500 years later, at the turn of the *Boreal/Older Atlantic*, the vegetation pattern of Switzerland changed quite fundamentally. Instead of lowland pine forests, mixed oak forests with hazel and silver birch were now widespread in the colline vegetation belt, and in the montane belt, deciduous forests with silver fir, which had migrated from SE and SW in the Central and eastern Northern Alps. In the subalpine zone in the eastern and central part of the Grisons, the newly immigrated Norway spruce spread all over from the east. In the montane belt, silver fir immigrated from the SE and SW. In the forest border area of the Central Alps of Valais and the Grisons,

Fig. 10: Vegetation of Switzerland during the Early Preboreal (ca. 9,500 years BP). Burga & Perret, *Vegetation und Klima,* 1998.

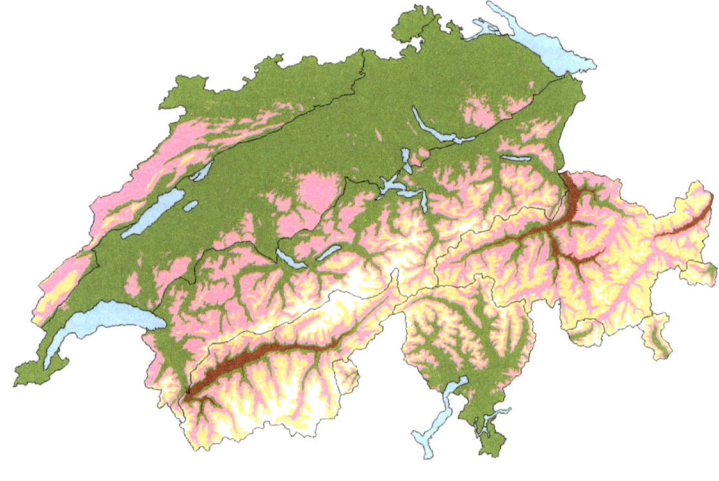

■ Pine-silver birch forest, locally juniper. In South-Ticino first mixed oak forests (oak, elm, lime, maple) and around 9,000 years BP immigration of silver fir.
■ Pine forest with silver birch and juniper. In the Central and Southern Alps Swiss stone pine, European larch, and dwarf shrub heath. In the eastern Pre-Alps, Swiss stone pine.
■ Alpine meadows and scree vegetation, rock vegetation.
■ Scots pine, dry meadows, and steppe.
☐ Ice and snow.

Fig. 11: Vegetation of Switzerland at the turn of the Boreal/Older Atlantic (ca. 8,000–7,500 years BP). Burga & Perret, *Vegetation und Klima,* 1998.

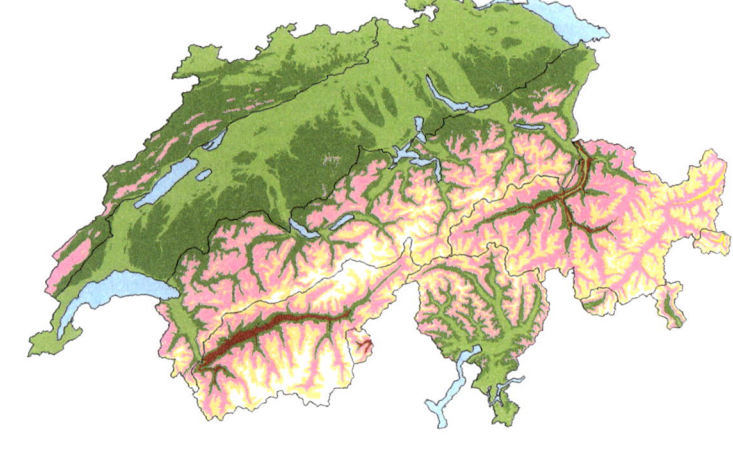

■ **Colline belt:** mixed oak forest, widespread hazel, Scots pine, silver birch. **Southern Alps:** mixed oak forest, silver birch, pine, some hazel, widespread alder. Hop hornbeam and manna ash.
■ **Montane belt:** hazel, mixed oak forest, immigration of silver fir from SE and SW in the Central and Eastern Pre-Alps, mixed deciduous woodland.
■ **Subalpine belt:** immigration of Norway spruce in the eastern and central parts of the Grisons. Silver birch and mountain pine, partly hazel. In the Central Alps mainly European larch-Swiss stone pine forest, dwarf shrub heath. In the Pre-Alps also elm and sycamore maple.
■ **Alpine belt:** alpine meadows and scree vegetation, rock vegetation.
☐ **Nival belt:** cryptogams, lichens, ice, and snow.
■ Scots pine, dry meadows, and steppe.
■ Scots pine, Norway spruce, and steppe.

Fig. 12: Vegetation of Switzerland during the Subboreal (ca. 5,000–2,500 years BP). Burga & Perret, *Vegetation und Klima,* 1998.

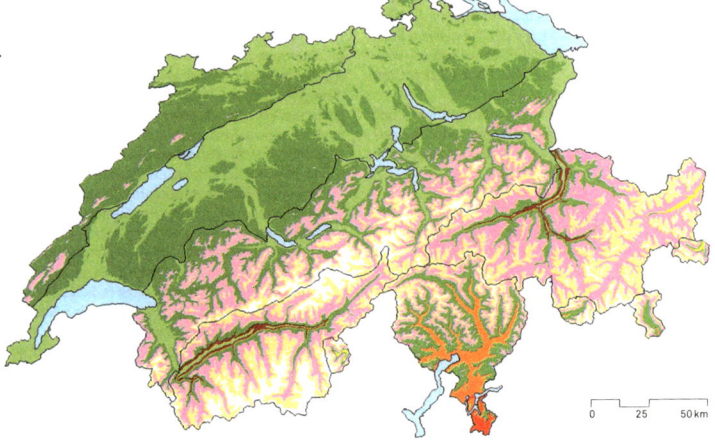

■ **Colline belt:** Northern Alps: beech forest, decreasing mixed oak forest. Yew, box, Scots pine, silver fir, silver birch, alder, hazel, hornbeam.
■ **Colline belt:** Southern Alps: oak-silver birch forest, beech, hornbeam, ash.
■ Hornbeam and hop hornbeam, manna ash, lime, oak, downy oak.
■ **Montane belt:** beech-silver fir forest and silver fir forest, mixed deciduous woodland, yew, partly Norway spruce forest. Hazel, alder, silver birch, pine.
■ **Subalpine belt:** Norway spruce forest, silver birch, Scots and mountain pine, green alder, juniper. Uppermost: European larch-Swiss stone pine forest (Central Alps), dwarf shrub heath.
■ **Alpine belt:** alpine meadows and scree vegetation, rock vegetation.
☐ **Nival belt:** cryptogams, lichens, ice, and snow.
■ Scots pine, steppe, and downy oak.
■ Scots pine, Norway spruce, and steppe.

0   25   50 km

European larch-Swiss stone pine forests grew. The formerly extensive alpine belt was now constricted from below by the ascending Norway spruce forests (Fig. 11). During the *Younger Atlantic*, beech immigrated into north-eastern Switzerland in the colline and montane belts around 6,000 years BP.

About 1,000 years later, in the *Subboreal*, the vegetation of Switzerland changed once again quite strongly, as the immigrated beech became widely dispersed on the Central Plateau and in the Northern Alps, at the expense of the existing mixed oak forest. In addition, silver fir played an important role in montane mixed forests, where silver fir-beech and purely silver-fir forests, partly Norway spruce forests, grew. In the subalpine belt, Norway spruce dominated, whose forests reached their Holocene optimum. The changes in Ticino were striking, where the oak-birch forest on acid soils became established in the colline belt, the beech in the colline and montane belts, and in southern Ticino the most thermophilic forest community in Switzerland – the manna flowering ash and hop hornbeam forest (*Fraxino orni-Ostryetum*) with downy oak (*Quercus pubescens*) (Fig. 12).[15] Later, due to human land appropriation, the oak-birch forest of the gneiss and granite zone was largely replaced by the introduction of sweet chestnut (*Castanea sativa*) since around Roman times.[16]

Summary

The Late-Glacial and Post-Glacial immigration and dispersal of important tree species in Switzerland is based on the first pollen-analytical database of Switzerland for the last 250,000 years, compiled by the author in 1989–1994, taking into account 618 peat bogs and lake sites.

About 13,000–12,000 years ago, the oldest coniferous forest communities in Switzerland started to establish themselves with the first European larch-Swiss stone pine and pine forests, after these tree species returned from their glacial refugia. From the beginning of the Holocene (ca. 11,000 years BP), the immigration and dispersal of thermophilic trees and shrubs, such as oak, lime, elm, maple, ash, hazel, and alder, later Norway

15 Ivo Ceschi, *Il bosco del Cantone Ticino,* Dipartimento del territorio. Divisione dell'ambiente, Armando Dadò, Locarno 2006.
16 See note 1.

spruce, silver fir, and beech began. From about 7,000 years BP, an increasing anthropogenic influence of the vegetation becomes apparent (forest clearings, pasture farming, and lowering of the Alpine timberline).

The Swiss stone pine refugia of the Last Glacial (LG, "Würm Glacial") were located at the southern edge of the Alps and in the southeastern Alpine region, plus in the Carpathians. Already in the Late-Glacial, the two Swiss stone pine sub-areas of the Pennine and Rhaetian Alps were formed. In the Southern Alps, European larch-Swiss stone pine forests grew from about 12,000 years BP at 1,700 m, reaching from the Gotthard southwards to the Camoghè-Tamaro-Limidario line. In the Central Alps of the Grisons and eastern Pre-Alps, Swiss stone pines appeared already around 13,000 years BP; into Valais and the Western Alps, the Swiss stone pine immigrated later, indicative of an east-west migration path. Climate and topography are cited as the cause of the two Swiss stone-pine areas. The glacial refugia of the Norway spruce were in western and central Russia, in the Carpathians, in the Dinaric Mountains, and on the eastern edge of the Alps. Around 9,000 years BP, Norway spruce migrated to the upper Adige Valley and further northwest to Engadine, which it reached around 8,000–7,500 years BP. Around 7,500–6,500 years BP, it reached, from the Vorarlberg and Rhine Valley, the Vorderrhein and Hinterrhein valleys, the Oberhalbstein and, via the San Bernardino Pass, Misox and northern Ticino. About 5,500 years BP, Norway spruce spread westward to the Lake Constance/Lower Lake Constance-Zurich-Entlebuch-Thun-Simmental line. At the same time, the Norway spruce reached the Upper Valais from SE or SW via the Simplon Pass and Lower Valais over the Forclaz Pass. It reached SW Jura around 5,500 years BP, which is confirmed by genetic studies of Norway spruce stands. The glacial refugia of the silver fir were in the southern Balkans, Anatolia, and the South Apennines. Around 9,000 years BP, the silver fir reached southern Ticino via two immigration branches from SE (Bergamasque to Venetian Alps) and SW (Liguria to Piedmontese Alps). From here, it rapidly migrated northward via Leventina and Misox. Around 7,000 years BP, its dispersal front reached the Prättigau-Appenzell-

Zurich Oberland-Rigi-Simmental line, and around 6,000 years BP, the Central Plateau and Central Jura were reached. Genetic studies of silver-fir stands confirm these migration routes.

An overview provides the migration paths and altitudinal distribution of forest-forming trees and the formation of the present vegetation belts of Switzerland. Around 13,000 years BP, the Late-Glacial reforestation with silver birch, Scots and mountain pine and, around 12,000 years BP, increasingly with Swiss stone pine and European larch with a timberline up to 1,700 m started. Thanks to a rapid temperature increase in the Preboreal, the timber-line rose to about 2,300 m, naturally fluctuating by 50–180 altitude meters. In the course of the middle Holocene, thermophilic hardwoods (oak, elm, lime, maple, and ash) along with hazel and alder migrated into the colline and montane belts, and silver fir and Norway spruce along with green alder into the montane and subalpine belts. The natural vegetation of Switzerland changed quite fundamentally several times during the last 11,000 years: The Preboreal birch-pine forests of the Central Plateau were replaced by mixed oak forests with hazel and silver fir in the montane belt at the turn of the Boreal/Older Atlantic, while in the subalpine belt of eastern Switzerland, Norway spruce was added to the already existing conifers.

Around 6,000 years BP, beech migrated into Northeastern Switzerland from the Lake Constance area. About 1,000 years later, in the Subboreal, the vegetation changed once again, as in the Central Plateau and in the Northern Alps the beech dispersed strongly and in the subalpine level the Norway spruce forests dominated. From about the Neolithic onwards, the forest area was reduced and the upper timberlines were lowered by several hundred meters in altitude compared to the potential natural upper limit as a result of the reclamation of arable land and clearing of alpine pastures. This marked the beginning of the change from a natural landscape to a landscape cultured by settled man.

Landscape. A Dialogue

Thomas Kissling

Walking, whether strolling through a city or hiking in the Alps, thrives on the continuous change of one's plane of reference, which hones one's awareness. The immediate view of the surroundings of one's feet gives way, in the next moment, to looking towards the horizon. And while what is immediately in front of us can be precisely named and classified, the contours of the individual objects become increasingly blurred as we look into the depth of space. The individual parts of the vista now flow into each other and turn into – a landscape.

However, Lucius Burckhardt teaches us that landscape as such is a construct, a concept. And that in a double sense. On the one hand, it is defined as the work of man, who manipulates and shapes. At the same time however as part of our imagination, which is shaped by culture and the knowledge we continuously acquire by reading books on natural science, by advertisements and films but also in the course of countless walks through actual landscapes.

This "accumulation" of details, of which we become aware, determines our perception. Consciously looked at, any landscape is broken down into layers and slowly decoded. Thus, different aspects and meanings come to the surface and are set in relation to each other. What is needed, in addition to an attentive viewer, is time and the conceding of an individual visual experience. This results in different readings of the landscape: while passing clouds, which draw a play of light and shadow on the meadows and ponds below, can be read as harbingers of the weather to come, pools of water in the fields refer to a thunderstorm the day before. Or they can be interpreted as an indication of highly compacted soil, no longer able to absorb any further moisture.

But perception is not limited to sight, for a landscape communicates itself to us in a variety of ways. The body acts as a kind of instrument, equipped with a sophisticated sensorium. When we hike in the mountains, for example, we listen to the crunch of gravel under our feet, smell the scent of a freshly mown meadow, feel the cold of the morning air and the heat of the afternoon, or taste the fresh water of a mountain stream. Or we are dazzled by the glistening sunlight reflected by the last snowfields between high rock tors. Individual aspects always appear more pronounced than others. If this is limited to a specific period of time, our perception will change quite fundamentally. On an early summer walk in Munich or Berlin, the scent of blooming linden trees is enchanting, revealing the very promise of a landscape to us – and not only on the famous street appropriately named "Unter den Linden" either. Rather, the scent appears to transform the whole city.

This many-faceted space cannot possibly be portrayed in an exhibition. Not least because the experience here is strongly visual. The afore-mentioned oscillation of seeing is not possible to the same extent as in a real landscape though, as the scale is different. The architectural space has a delimiting, sometimes even constricting effect here. Even freely walking gives way to guided movements or, at least, is ruled by an orderly choreography. On paths, in cul-de-sacs, through passages, and across intersections, we move from the entrance to the exit and are guided past exhibits that, in most cases, are catalogued and explained.

Often, the path through a landscape is not absolutely fixed but must be sought and found. On the one hand, our movements are determined by the elements we find. A thorny undergrowth may force me to walk around a wood. A steep slope may shape my originally linear walk into zigzagging along the contour lines. The course of a stream makes me search for some stepping stones to cross the water on dry feet. On the other hand, all these conditions have a determining effect. A slippery meadow will cause me to choose each step carefully, slowing my progress, and frequently force me to an absolute standstill.

To transform a landscape into an exhibition requires a translation, a careful process of subtraction. Much like an archaeologist who uses a digging hammer, a round brush, or a tongue trowel to work his way forward, layer by layer, in order to gain access and uncover the "core" of a place. However, the goal of the search cannot be clearly stated in advance. At most, it can be sketched out. And so the procedure is characterized by an attentive gaze. For the object at which the search is aimed could, the very next moment, be damaged or even irrevocably destroyed by a wrong move. This care in searching or seeking out an object embedded in a landscape is not relevant for tourists though. Their attention is focused on the places declared to be "beautiful" by the broad masses. Their visits and consumption, however, in the end irrevocably shatter these landscapes of their longing.

The procedure of digging or excavating with a view to determining the essence of a landscape, however, is not about striving for objectivity or schematic clarity by means of a reduction. Rather, the operation is deliberately evaluative from the author's perspective. This with the aim of identifying the substrate for a simulation capable of creating an immediate landscape experience in the exhibition, opening the door to the viewers' imagination. This is the only way to render the desired richness of perception possible.

Migrating Landscapes

Günther Vogt, Violeta Burckhardt, and Simon Kroll
(Case Studio VOGT), Arsenale, Venice

Photos by Alessandra Chemollo

At first glance, the relief composed of pressed bricks is reminiscent of a classical architectural model. Only on closer inspection, the supposed *city en miniature* is revealed to be a fragile structure, whose potential lies in the possibility of change. This is literally incorporated into the individual stones.

The soils from which the individual parts are made are from different regions of the Alps and the rest of Europe. Together, they form a kind of mapping of the diverse pedological situations on the continent. Embedded into the conglomerate are seeds of a wide variety of plants that are considered "non-native" to the urban landscape of Venice, but have been continuously introduced to the port city over the past centuries. Venice's role as one of the most important trade hubs between Western Europe and the Eastern Mediterranean provided the basis for the establishment of a broad spectrum of vegetation that enriches the city and constantly continues to expand to this day.

Even the mature plane trees that surround the relief are not native to the area. Nevertheless, they form one of the most frequently planted urban tree species not only in Venice but throughout all of Europe. For that matter, the plane tree is often found along rivers: for example, the Thames in London, the Seine in Paris, or the Tiber in Rome. This is primarily due to site conditions. The humid gravel plains of river valleys in Greece, Albania, and Croatia, the origin of the plane trees, are comparable to the conditions of our cities. Sealed surfaces and increasingly heated outdoor spaces call for an adapted vegetation that decisively renounces the ideologically motivated distinction between native and invasive.

"Urban nature" turns into a decisive factor when we talk about the future of urban spaces, in which controlling certain processes is increasingly difficult. This "loss of control" is not only evident in the built environment, but also in the landscapes, where invasive plants, for example, are beginning to encroach on the local forests, permanently changing not only the ecosystems but also our perception.

The relief picks up on the theme of our loss of control. Once initiated, the process – an attempt in the true sense of the word, in which failure is a conceivable option – cannot finally be controlled either. Only a single tendency can be specified: its thriving vegetation will successively dissolve the precise form over the course of the Biennial. The morphosis from "mineral" to "organismic" will be driven by the uncontrollable forces of rain, wind, and sun: creating a landscape in a constant state of becoming.

The Vegetative Nervous System

Gerda Steiner & Jörg Lenzlinger

Gerda Steiner & Jörg Lenzlinger create a magical but wild and uncontrolled growth, stimulating our curiosity, amusing and striking amalgams of the organic and the inorganic; installative atmospheres in which interactions and unpredictable coexistences may occur. The principle of preservation (of the found, marginal, and perhaps unappreciated) is always accompanied by a tendency to organically transform form and synthesis. Chance, playfulness and plan, animistic and alchemistic strategies intertwine.

Spectators are invited to immerse themselves in a fantastic vegetative universe or to participate in experiments that awaken their senses and minds. The artists' installations, as disquieting as they are enchanting, forge connections between antagonistic worlds, proposing to observe the strange laboratory of all things alive with its biodiversity and to question notions of fertility and growth. They argue for poetically inspiring symbioses, mixtures rather than monocultures, systemic interactions and taking seriously the beauty of complexity and diversity.

Text by Kirsten Claudia Voigt

Cartography of Hidden Landscapes

Type and Profile

*This was the order of human institutions: first the forests,
after that the huts, then the villages, next the cities,
and finally the academies.*

Giambattista Vico, *The New Science*, 1725

Landscapes not only provided the setting but rather the
very substrate for the development of human and,
later, urban culture. The chronological development from
nature to cultivated landscapes to urban landscapes
and townscapes is akin to geomorphological processes:
depositing, layering, folding, distorting, and eroding.

Originally shapeless clay is transformed by means of
pressure and heat into an endless number of consistently
identical components. Lime, water, gravel, and sand, end-
lessly and almost arbitrarily shaped, solidify into concrete.
They are the enduring parts of the built urban conglomerate – just like the façades of Haussmann's urban
renewal in Paris. Depending on their location within
the architectural composition, they reflect different
qualities of stone quarried in the landscape, as it was then,
in and around Paris. Like a transect through the original
landscape. A silhouette of itself. Richly decorated stuccoes
of the interiors testify to the underground extraction
of gypsum. The resulting caves or caverns were subsequently used for the cultivation of edible mushrooms. The
geology of the city ultimately consisted of the reshaping
of the land with the resources of that very landscape.

Even the urban landscape – the urbanized landscape,
that is – the transformation processes, from liquid to solid,

deposits to layers, recall geological processes, although in quite different time dimensions. Where do we find this landscape today though? Let's trace it. Today, the original landscape is more often than not hidden deep underneath the city and only visible as a "watermark" in the morphology of the built city.

In London, the Thames cut a gentle river valley into the flat landscape. The infrastructural buildings respected any resistances contingent on the topography. The first railroad lines followed the contour lines, creating centers. Industrial buildings were erected along the river because of their dependence on the electrical energy produced there. To any attentive pedestrian strolling through the city, the finely river-crafted topography is still apparent to this day though.

In Paris, urban parks such as Buttes Chaumont are evidence of the landscapes shaped by the original use of the land as limestone quarry and part of the vocabulary of urban-planning: perimeter developments, ribbon developments, and solitary townhouses but also squares, promenades, or cemeteries. They all follow the new rules of urban typology.

On the Nature of the City. From City to Landscape

The original pair of opposites, consisting of a city versus a landscape, has dissolved. The question now is, in what way landscapes have been urbanized. To think and define the city from the landscape implies thinking, designing, and developing a profile of landscape as such. What does this mean? Unlike the urban type set, the profile of a landscape is more broadly defined in time and space. Any profiling landscapes operates on the basis of surfaces and cross sections, with visible and perceivable differences: from urban neighborhood to regional landscape, from urban park to forest, and from horticulture to agriculture.

In turn, the Alps as a superordinate landscape figure and complex space of reference for peripheral metropolises are part of the landscape of Central Europe. Eight countries, four languages, countless dialects, and yet one and the same cultural space. The geographic space is fragmented into an infinite number of landscapes. Thus a single profile of the Alpine landscape in it entirety must needs fail. A wealth of individual landscape profiles helps solve this dilemma. The development of the metropolises depended on topography, hydrology, geology, and climate and generated diverse relationships with the surrounding landscapes, which can be read in the very culture of a city. While in Munich, the Alpine region is a remote hinterland, there is nonetheless a strong reference to be found in its restaurants, cuisine and also in its fashion. In Marseille, however, this seems to play a much less important role, for although Alpine geology is even visible close to the downtown area, the urban culture negates this landscape. In fact, the Mediterranean is its gravitational space, Northern Africa its reference space, whilst the topographical situation that prepares the ground for this orientation is partly the result of the very genesis of the Alps. The inclined plane guides one's view forward, into the open and wide expanse of the sea, away from the craggy rocks and steep ridges at the rear of the city.

Obviously, landscapes always have many different authors and interpreters.

Günther Vogt

> Urban ring: The Alps are surrounded by metropolitan areas and large urban regions. The red line marks the Alpine Convention © Chair of Günther Vogt, ETH Zurich. Input data: geoVITe, Alpine Convention (GIS ALPARC, 2008). Using QGIS Geographic Information System.

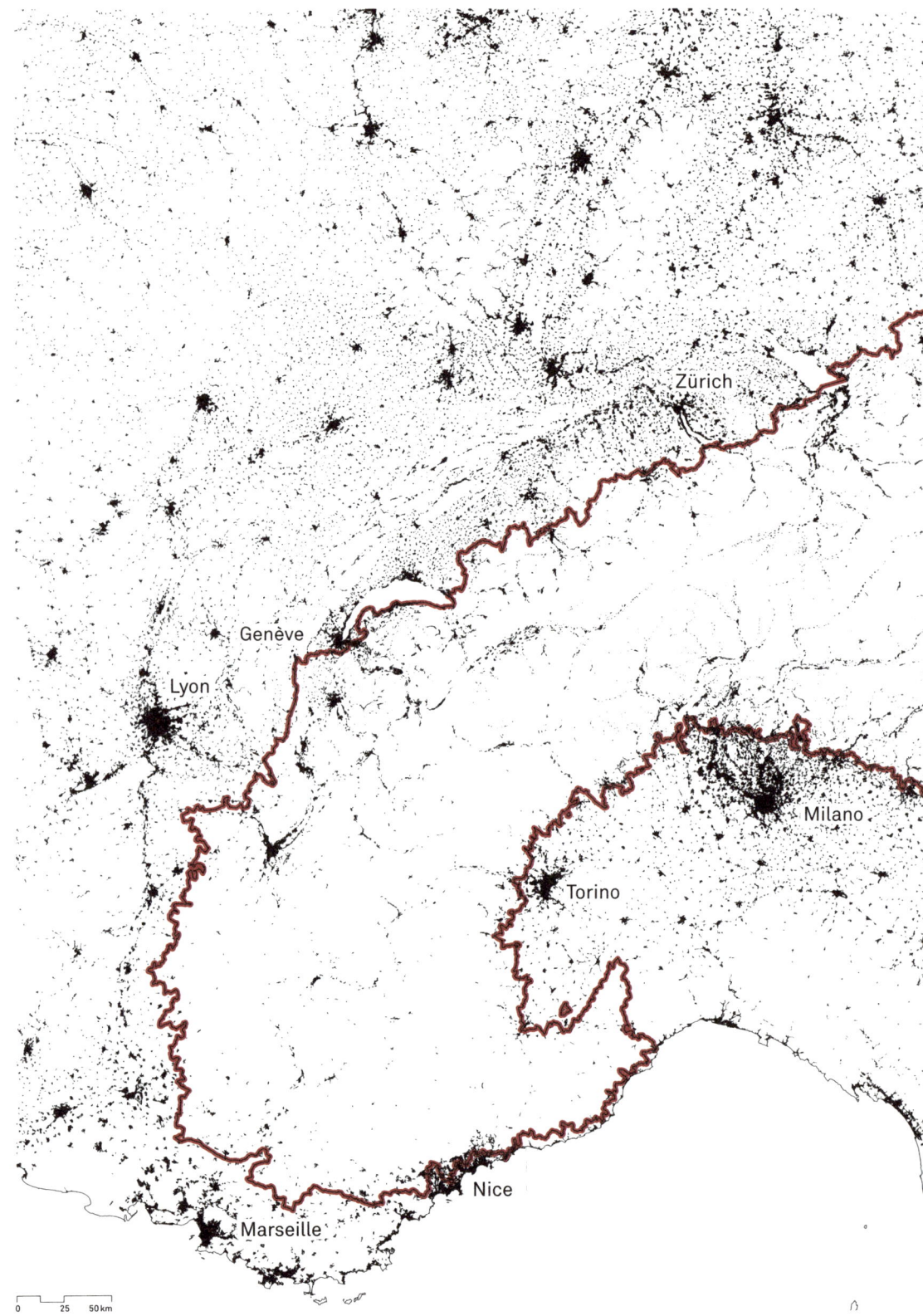

Zürich

Genève

Lyon

Milano

Torino

Nice

Marseille

0   25   50 km

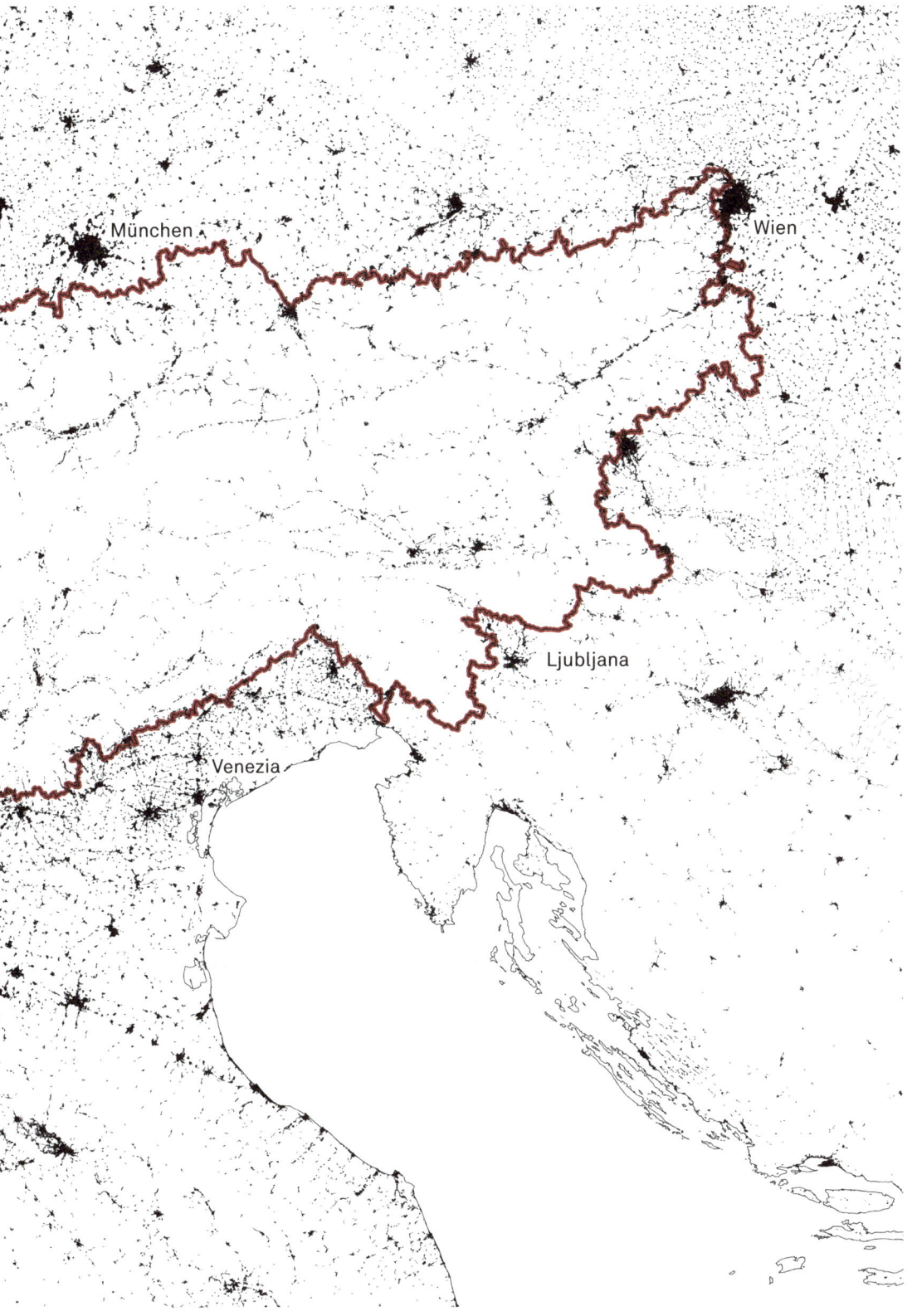

München

Wien

Ljubljana

Venezia

Amalia Bonsack (*1991) studied Architecture at the Swiss Federal Institute of Technology in Lausanne (EPFL) and Zurich (ETH), where she graduated in 2017 having specialized in Urban Design. She has gained a range of professional experience in Switzerland and abroad and is a founding member of the architecture collective la–clique. Since 2018, she has been a teaching and research assistant at the Chair of Prof. Günther Vogt at ETH Zurich. Besides her involvement as Edition Manager of the book *Mutation and Morphosis*, she is in charge of the Territory of the City course, which deals with current transformation processes in European metropolitan landscapes.

Maren Brakebusch (*1974) studied Landscape Architecture at the Hanover Leibnitz University, was Project Manager at the office of Kamel Louafi Landscape Architects for various projects within the scope of EXPO. Since 2002, she has worked as Project Manager and since 2009 as Office Manager for VOGT Landscape Architects in Zurich. In 2014, she took over the overall office management of VOGT and became a member of the VOGT Board of Directors, which she has chaired since 2017. In addition to lectures and jury activities, she has been teaching students of architecture at FH Potsdam since 2017 and, since 2020, at ETH Zurich as part of a one-year lectureship.

Violeta Burckhardt (*1987) studied Architecture in Mexico City (UNAM) and Urban Design at TU Berlin and Tongji University in Shanghai. She is currently working with VOGT Landscape Architects in Zurich, where she is developing research and practice-oriented projects as part of the VOGT Case Studio. Based on her experience in film and art production, she has curated several cultural events and co-designed several exhibitions as a member of her office. She recently finished the production and organization of VOGT Landscape Architects' contribution to the "How will we live together?," Biennale di Venezia – a process piece exploring the shifting of ecosystems as a result of human activities – and is member of the Chair of Günther Vogt at ETH Zurich. Together with Günther Vogt, she is co-author of the book *Paradise Now. Die neuen Grenzen des Gartens* (published by Matthes & Seitz, Berlin, 2021), a series of small essays reflecting on the limits of the garden today by looking at the world as an artificial paradise.

Conradin A. Burga (*1948) studied 1970–1976 geography, geology, petrography, and chemistry at the University and ETH Zurich (Diploma 1976) and botany at the University of Basel (PhD 1980). After a postdoc as research associate at the Botany School (now Dept. of Plant Sciences) of the University of Cambridge (GB), he joined the Department of Geography of the University of Zurich as senior research assistant and 1996–2013 as Professor of Physical Geography/Biogeography. He is specialized in plant geography/biogeography and Quaternary palaeoecology, especially of the high mountain ecosystems of the European Alps (vegetation dynamics; floristic, vegetation, and climate history; palynology). He was Vice-President of the Hannover Tüxen Society of Vegetation Science, founder and head of Biomonitoring/Global Change, an international work group, and member of several national and international commissions. In addition, he was 1999–2013 the

Editor-in-Chief of the journal of the Society of Natural Sciences Zurich (founded in 1746). His research and teaching covered ice-age history and the Quaternary climate and vegetation history of the Swiss and Italian Alps, and various other biogeographical aspects such as the plant ecology of glacier forefields and vegetation mapping of the Swiss Alps.

Julian Charrière (*1987) is a French-Swiss artist based in Berlin, Germany. Working across media and conceptual paradigms, his projects often stem from fieldwork in remote locations with acute geophysical identities and combine unconventional methods to point out the inseparable bond between human civilization and the environment, revealing unexpected perspectives and addressing central issues of deep geological as well as human historical times. A former student of Olafur Eliasson and participant in the Institute for Spatial Experiments, he graduated from the Berlin University of the Arts in 2013 and has exhibited his work at institutions and biennials worldwide, among others including the Dallas Museum of Art (2021), Centre Pompidou (2019), 57th Biennale di Venezia (2017), and Antarctic Biennale (2017). (http://julian-charriere.net)

Alessandra Chemollo (*1963) graduated in 1995 from the IUAV University Institute of Architecture (IUAV) with a dissertation on the relationship between architecture and photography. Her reflections on representing architectural works and, more widely, on anthropological landscape crosses both her commissioned work and her autonomously produced projects, without a solution of continuity. She illustrated a number of monographs with her photographs, mastering specific reading skills on architectural manuscripts with her takes on documentations as a starting point. In more than thirty years of professional experience, her work has ranged from historical architecture to contemporary buildings and has developed theoretical circles with didactic and curatorial finality. Since 1986, she has been working as a photographer; from 1991 till 2014, she worked with Fulvio Orsenigo. Since 2013, she has been Adjunct Professor of Photography of the Landscape and Garden MA course at IUAV, Venice. (http://alessandrachemollo.it)

Sophie von Einsiedel (*1995) raised between Germany, Switzerland, and the United Kingdom, has read Geography at Girton College, (University of Cambridge), where she graduated with a double first in 2017. She then studied Spatial Design at Central St Martins in London. In 2020, she received her Master's degree in Landscape Architecture from the University of Edinburgh. Since joining VOGT Landscape Architects in Zurich in 2020, Sophie von Einsiedel has worked on the LIFE exhibition by Olafur Eliasson at Fondation Beyeler as well as on other conceptual and research projects for VOGT Landscape Architects and the Case Studio.

Thomas Kissling (*1980) studied architecture at ETH Zurich, having previously trained as a draftsman, graduating in 2010 under the tutelage of Prof. Wolfgang Schett. Since 2010, he has been a research assistant and, since 2016, a research associate at the Chair of Prof. Günther

Vogt at ETH Zurich. Since 2021, he has been part of VOGT Landscape Architects in Zurich. His teaching and research focus on transformation processes in the Alpine landscape and on the change in the meaning and use of public open spaces in metropolitan territories as a result of advancing urbanization. He is Co-Editor of *Landscape as a Cabinet of Curiosities* (2015), a compilation of five conversations with Günther Vogt. In addition, Thomas Kissling published *Mutation and Morphosis – Landscape as Aggregate* (2020) together with Günther Vogt. The book looks at numerous aspects involved in the collective process of exploring, designing, and shaping landscape, from planning to implementation, at the interface between academy and practice.

Andreas Klein (*1987) is an architect based in Zurich. After completing his studies at ETH Zurich in 2013, he worked for several architectural firms and undertook various study trips all over the world. In 2016, he started as a research and teaching assistant at the Chair of Prof. Günther Vogt at ETH Zurich. In addition, he works with architect and writer Tinetta Rauch on architectural projects of various scales, most of which in an Alpine context.

Simon Kroll (*1984) studied Landscape Architecture at the Hanover Leibnitz University from 2005 till 2010. He has been part of the VOGT Landscape Architects team in Zurich since October 2010. Since 2012, Kroll has worked as a project manager with a wide range of professional experience and a deep understanding of complex performance requirements. As Head of the VOGT Case Studio, he has been part of the VOGT Management since 2017. Since 2020, he has been a teaching assistant at the Visiting Studio of Landscape Architecture of Maren Brakebusch at ETH Zurich, focusing on urban landscapes and the effects of heat islands.

Max Leiß (*1982) is a visual artist specializing in sculpture and photography. He studied Fine Arts at AdBK Karlsruhe and ENSBA Paris and graduated in 2011. His international recognition includes the awards "junger westen" and the Kalinowski-Preis in 2017, nominations for the Swiss Art Awards in 2013 and 2017, and a residency at Cité des Arts, Paris, in 2014. His work was exhibited at institutions such as Kunsthaus Baselland, Kunsthaus Aarau, Kunsthalle Basel, Kunsthalle Recklinghausen, as well as at Nicolas Krupp Galerie, Galleria Enrico Fornello, and Galerie Meyer Riegger. Max Leiß was teaching and researching in the fields of art, landscape, architecture, and strollology at EPFL Lausanne in 2019 and since 2017 at ETH Zurich.

Katie Paterson (*1981) is known for her conceptual artworks that have involved live broadcasting of the sounds of a melting glacier, mapping all the dead stars in the universe, compiling a slide archive of the history of darkness across the ages, custom-making a light bulb to simulate the experience of moonlight, burying a nano-sized grain of sand deep within the Sahara Desert, and sending a re-cast meteorite back into space. Katie has exhibited internationally, from London to New York, Berlin to Seoul, and her works have been included in major exhibitions such as the Tate Britain, Hayward Gallery, Museum of Contemporary Art Sydney, Guggenheim Museum New York,

and Scottish National Gallery of Modern Art. She was the winner of the Visual Arts category of the South Bank Awards and is an Honorary Fellow of the University of Edinburgh. (www.katiepaterson.org)

Markus Ritter (*1954) grew up in Basel and has remained attached to his hometown throughout his entire life, though even while still a child, he longed for anything to do with nature. On his mother's side, he got to know the Walser people, full-bearded Alpine herdsmen living close to nature. Thus, the boy gloried in many forms of a pre-scientific experience of nature, from hunting snails to butterflies, meeting teachers who introduced him to nature in all its splendor. No wonder, his career was to coincide with these personal interests. Among other things, he works for the Schweizerische Vogelwarte, the Swiss Ornithological Institute in Sempach and the Naturschutzbund, the Nature Conservation Society in Basel. Nature in the city was one of his foremost interests, and he is a co-author of the Basler Natur-Atlas (1985), celebrating urban ecology. In addition, his life is filled with biological and ecological project work and lectures. In 1986, gloomy prospects for nature and the environment caused him to found the Green Party in Basel with Lucius Burckhardt and to enter into a highly varied co-operation with him. 1988–2001, he was a contributing member of the Cantonal Parliament. The Rio Conference of 1995 made him believe the ecological breakthrough had inexorably begun. In 1996, he and two other biologists founded a joint-stock company for ecological consultancy and services. His responsibilities have become quite extensive and comprise being on committees of associations, foundations, and boards of directors. As part of the executive government staff (2006–2018), for eight years in the Präsidialdepartement, the Cantonal Department of Presidential Affairs, he got to know the state in its most diverse facets. Now retired, he delves with relish into the topics of his own experience and of omnipresent nature around us.

Lukas Ryffel (*1992) born in Baden, at the foot of the Linth Glacier, grew up surrounded by a plant nursery his parents have in Uster, studied biology and architecture at ETH Zurich, the Imperial College London, and CEPT Ahmedabad. He is part of the 8000.agency collective, which deals with ongoing transformation processes in the city of Zurich, and teaches at the Chair of Jan de Vylder at ETH Zurich. Since 2021, has been working as a research assistant at the Chair of Günther Vogt at ETH Zurich, where he mainly investigates the Alpine water landscape and its transformation in the wake of climate change.

Roland Charles Shaw (*1984) read Oriental Studies (Sanskrit and Ancient Greek) at St. John's College (Oxford) and Ludwig Maximilian University (LMU Munich). In 2014, he completed a degree in architecture at the Architectural Association in London (AA Dipl.), where he was also co-editor of AArchitecture, a student magazine. His final project (Diploma Unit 4) dealt with potential agricultural scenarios in Alentejo, Portugal, and comprised an olfactory archive of the landscape and a film. He has been a research assistant at the Chair of Günther Vogt at ETH Zurich since 2015, where he is responsible for Urban Food, an elective course, and technical modules in GIS, animation, and film.

Gerda Steiner (*1967) & **Jörg Lenzlinger** (*1964) live in Switzerland. They participated in Expo02 with their gigantic "Heimatmaschine." At the 2003 Biennale di Venezia, they represented Switzerland with a delicate "Falling Garden" inside San Stae church and other invasive installations at the 21st Century Museum in Kanazawa, Japan, in the baroque Abbey Library in Saint Gall, at the Biennial of Contemporary Art in Seville, at the Moscow Biennial and, more recently, at Kunsthaus in Bregenz and Museum Tinguely in Basel. Their permanent installation "Vegetative Nervous System" is shown at Museum Kunstpalast in Düsseldorf. (www.steinerlenzlinger.ch)

Günther Vogt (*1957) Günther Vogt's training at Gartenbauschule Oeschberg provided the practical basis for his intensive landscape work. His knowledge of vegetation and his skills in cultivation continue to be the very cornerstone of his work. His studies with Peter Erni, Jürg Altherr, and Dieter Kienast at Interkantonales Technikum Rapperswil combined the disciplines of culture, design, and natural sciences. VOGT Landscape Architects emerged from the office partnership with Dieter Kienast in 2000. With projects such as the Tate Modern in London, the Allianz Arena in Munich, or the Masoala Rainforest Hall at the Zurich Zoo, the firm has achieved international recognition. Its work is characterized by the dialogue established between the various disciplines and its close cooperation with artists. Since 2005, Günther Vogt has been pursuing a combination of teaching, practice, and research with his chair at the Institute of Landscape Architecture at ETH Zurich. As a passionate collector and keen traveler, he is looking for ways to read, interpret, and describe landscapes, and finding answers to questions about future forms of urban coexistence. In 2012, Günther Vogt was awarded the Prix Meret Oppenheim by the Federal Office of Culture.

Gerhart Wagner (*1920) grew up in the rural area of Bolligen near Bern and his first hobby was ornithology. Hence, he studied Biology, Physics, and Geology in Bern and Geneva. In 1946, he obtained a grammar-school teacher's diploma, in 1949 his doctoral degree (diploma) with a dissertation on zoology. 1949–1950 he worked as a secondary-school teacher in Grindelwald, 1950–1958 as a grammar-school teacher in Bern. In 1956, he went on expedition to the Norwegian bird island of Røst in the Lofoten Islands. Concerned about the rapidly increasing radioactivity of the biosphere due to dozens of atomic explosions, he was elected by the Federal Council in 1958 to head the new Section of Protection Against Radiation and entrusted with the task of establishing the Radiation Protection Ordinance. In 1964, after its completion, he accepted a call from the University of Zurich as Assistant Professor of Zoology. In 1969, he returned to Bern as Principal of the newly established Neufeld Grammar School. A common thread running through his curriculum vitae was the collection of flowering and fern plants for his herbarium. This work (of approx. 12,000 sheets) is now permanently stored at the Conservatoire et Jardin botaniques de la Ville de Genève. After his retirement, and in collaboration with Konrad Lauber, *Flora des Kantons Bern* [*Flora of the Canton of Bern*] (1991) was published, followed by the first edition of *Flora Helvetica* (1996). His intensive work on glacial structures began with the unravelling of a particular moraine structure in the area where he lived. In 1996, the University of Bern awarded him an honorary doctorate for his contributions to zoology, botany, and geology.

Rolf Weingartner (*1954) studied, received his PhD, and habilitated at the Institute of Geography of the University of Bern, specializing in Hydrology. After some extended stays abroad, in Germany and New Zealand, he headed the Hydrology Group at the Bern Institute of Geography (1989–2019), first as Senior Assistant, then as Adjunct Professor and, finally, as Full Professor. Currently, he is co-owner of ecosfera gmbh, a company invested in the fields of water, forests, and ecosystems. He also holds various mandates in teaching, research, and administration. The topic of water not only includes many aspects that go beyond hydrology itself but also has great practical relevance. As a geographer by training, Rolf Weingartner feels very much connected to this interdisciplinary and applied perspective. His areas of expertise include water in Switzerland and, in this context, especially water resources and natural hazards given climate change. In this, he can also benefit from his more than thirty years of experience as Project Manager of the *Hydrological Atlas of Switzerland*. Another priority of his work is Earth's mountain regions with special focus on the Himalayas and the European Alps. His management of various major research projects and his chairmanship of the *Mountain Research Initiative* opened up various approaches to mountain issues to him.

SOLID FLUID BIOTIC
Changing
Alpine Landscapes

Edited by Thomas Kissling

Translations, copyediting and proofreading: Suzanne Leu
Coordination: Marius Wenger
Design: Integral Lars Müller/Lars Müller and
Esther Butterworth
Production: Esther Butterworth
Photography (Biennial contributions):
Alessandra Chemollo
Lithography: prints professional, Berlin, Germany
Printing and binding: DZA Druckerei zu Altenburg, Germany
Paper: Munken Polar 1.13, 130 gsm

Lars Müller Publishers is supported by the Swiss Federal
Office of Culture with a structural contribution for the years
2021–2024.

Lars Müller Publishers
Zurich, Switzerland
www.lars-mueller-publishers.com

ISBN 978-3-03778-677-2, English
ISBN 978-3-03778-690-1, German

Distributed in North America by ARTBOOK | D.A.P.
www.artbook.com

Printed in Germany

**Publications by and about Günther Vogt
available in our program**

Mutation and Morphosis
Edited by Günther Vogt, Thomas Kissling
2020, ISBN 978-3-03778-618-5, English
2020, ISBN 978-3-03778-619-2, German

Mara Katherine Smaby, Nicola Eiffer,
Nicole la Hausse de Lalouvière
Wunderlust/Wanderkammer
Edited by Günther Vogt
2016, ISBN 978-3-03778-489-1, English

Günther Vogt
Landscape as a Cabinet of Curiosities
Edited by Rebecca Bornhauser, Thomas Kissling,
Chair of Günther Vogt, ETH Zurich
2015, ISBN 978-3-03778-303-0, German
2015, ISBN 978-3-03778-304-7, English

Günther Vogt
Miniature and Panorama
2012, ISBN 978-3-03778-233-0, English

Franziska Bark Hagen
Versuche das Glück im Garten zu finden
Edited by Chair of Günther Vogt, ETH Zurich
2011, ISBN 978-3-03778-247-7, German

Jürgen Krusche
Strassenräume Berlin Shanghai Tokyo Zürich
Edited by Chair of Günther Vogt, ETH Zurich
2011, ISBN 978-3-03778-248-4, German

With the support of:

VOGT

ETH *zürich*

D ARCH

ERNST GÖHNER STIFTUNG

Image captions and sources:

Dust jacket Input data: geoVITe, using QGIS Geographic Information System

Book cover Julian Charrière, *The Blue Fossil Entropic Stories,* 2013 (copyright the artist; © 2021, ProLitteris, Zurich)

Page 11 © Leonard von Matt / Fotostiftung Schweiz [Swiss Foundation for Photography]

12/13 © ETH Picture Library, Zurich / photographer: Peter Hahn

14 top © ETH Picture Library, Zurich

14 bottom © ETH Picture Library, Zurich / photographer: Elektro-Watt (Zurich)

15–17 © ETH Picture Library, Zurich / photographer: Comet Photo AG (Zurich)

48, 49, 52–55, 134–139, 172–175, 178–189 © Alessandra Chemollo

77–83 Katie Paterson, *Fossil Necklace,* 2013 @ Katie Paterson

102–111, 114–117 Julian Charrière, *Towards No Earthly Pole,* 2019 (copyright the artist; © 2021, ProLitteris, Zurich)

118–131 Maps © swisstopo (edited by Lukas Ryffel), photos: © Lukas Ryffel

142, 143, 146–151 Photos © VOGT Landscape Architects

144 Postcard, center, right © Photoglob Zurich

145 Poster, Funicolare Monte Brè, Lugano, Switzerland, 1946, Otto Ernst ©, Museum für Gestaltung Zürich, ZHdK, Plakatsammlung
Poster, Lugano, 1952, unknown artist © Museum für Gestaltung Zürich, ZHdK, Plakatsammlung

191 Gerda Steiner & Jörg Lenzlinger, *Paysage au grand galop,* detail, Musée de Valence, 2013 © Gerda Steiner & Jörg Lenzlinger

192, 193 Gerda Steiner & Jörg Lenzlinger, *Brainforest,* 21st Century Museum, Kanazawa, Japan, 2004 © Gerda Steiner & Jörg Lenzlinger

194, 195 Gerda Steiner & Jörg Lenzlinger, *Les pierres & le printemps,* Domaine de Chaumont sur Loire, 2015 © Gerda Steiner & Jörg Lenzlinger

196, 197 Gerda Steiner & Jörg Lenzlinger, *Nationalpark,* Bündner Kunstmuseum Chur, 2013 © Gerda Steiner & Jörg Lenzlinger

We sincerely thank

the Department of Architecture (D-ARCH) of ETH Zurich, Ernst Göhner Stiftung and VOGT Landscape Architects for their welcome support,

all authors and artists for their contributions,

Li Fei, Claudia Gebert, David Jung, Ramona Köchli, Samira Lenzin, Lukas Ryffel, Sabine Sarwa and Edoardo Signori for their contributions and research support,

Suzanne Leu for her careful editing and precise translation,

Lars Müller and the team at Lars Müller Publishers (namely Esther Butterworth, Maya Rüegg, and Marius Wenger) for their dedicated and cordial collaboration in the development and realization of the book,

Günther Vogt for countless discussions, his goodwill, and the trust he has placed in us.